QUANTUM WALKS FOR COMPUTER SCIENTISTS

Synthesis Lectures on Quantum Computing

Editors

Marco Lanzagorta, *ITT Corporation*

Jeffrey Uhlmann, *University of Missouri, Columbia*

Synthesis Lectures On Quantum Computing presents works on research and development in quantum information and computation for an audience of professional developers, researchers and advanced students. Topics include Quantum Algorithms, Quantum Information Theory, Error Correction Protocols, Fault Tolerant Quantum Computing, the Physical Realizations of Quantum Computers, Quantum Cryptography, and others of interest.

QUANTUM WALKS FOR COMPUTER SCIENTISTS
Salvador Elías Venegas-Andraca
2008

Quantum Walks for Computer Scientists
Salvador Elías Venegas-Andraca

ISBN: 978-3-031-01383-6   paperback
ISBN: 978-3-031-02511-2   ebook

DOI 10.1007/978-3-031-02511-2

A Publication in the Springer series
SYNTHESIS LECTURES ON QUANTUM COMPUTING: #1

Series Editor: Marco Lanzagorta, ITT Corporation and Jeffrey Uhlmann, University of Missouri, Columbia

Library of Congress Cataloging-in-Publication Data

Series ISSN:
Series ISSN:

QUANTUM WALKS FOR COMPUTER SCIENTISTS

Salvador Elías Venegas-Andraca
Quantum Information Processing Group,
Tecnológico de Monterrey Campus Estado de México,
México

SYNTHESIS LECTURES ON QUANTUM COMPUTING: #1

ABSTRACT

Quantum computation, one of the latest joint ventures between physics and the theory of computation, is a scientific field whose main goals include the development of hardware and algorithms based on the quantum mechanical properties of those physical systems used to implement such algorithms.

Solving difficult tasks (for example, the Satisfiability Problem and other NP-complete problems) requires the development of sophisticated algorithms, many of which employ stochastic processes as their mathematical basis.

Discrete random walks are a popular choice among those stochastic processes.

Inspired on the success of discrete random walks in algorithm development, quantum walks, an emerging field of quantum computation, is a generalization of random walks into the quantum mechanical world.

The purpose of this lecture is to provide a concise yet comprehensive introduction to quantum walks.

KEYWORDS

Quantum walks, quantum algorithms, quantum information science, quantum computation, unconventional models of computation.

This work is dedicated to my beloved

D.A.R.

Godsons Cristóbal and Rafael.

Amparo, Samy, Bernardo, Margarita, and Humberto.

Future generations of quantum engineers and quantum programmers.

Contents

Preface

Quantum computation, one of the latest joint ventures between physics and the theory of computation, is a scientific field whose main goals include the development of hardware and algorithms based on the quantum mechanical properties of those physical systems used to implement such algorithms.

Solving difficult tasks (for example, the Satisfiability Problem and other NP-complete problems) requires the development of sophisticated algorithms, many of which employ stochastic processes as their mathematical basis.

Discrete random walks are a popular choice among those stochastic processes.

Inspired on the success of discrete random walks in algorithm development, quantum walks, an emerging field of quantum computation, is a generalization of random walks into the quantum mechanical world.

The purpose of this lecture is to provide a concise yet comprehensive introduction to quantum walks.

In addition to the Introduction chapter itself, our book starts with the three following chapters:

Chapter two introduces the postulates of quantum mechanics in a form suitable to be employed by computer scientists. By means of the state space postulate we present to the reader the quantum counterpart of the bit, the qubit. Moreover, the evolution postulate provides the mathematical structure that allows us to describe the behavior of a qubit as time passes by, while the measurement postulate presents the mathematical procedure to be employed in order to compute the probability distributions inherent to any computation based on quantum mechanics. We then introduce the most surprising (for a computer scientist) no-cloning theorem, which states that it is impossible to make perfect copies of arbitrary quantum states, and finish this chapter by delivering the rules for working with two and more qubits, and by introducing the concept of quantum entanglement as a computational resource.

Chapter three provides some essential elements of the theory of computation necessary to show and quantify the role of discrete random walks in computer science. We particularly focus on classical and quantum versions of Turing machines as well as on the definitions and theorems of the theory of complexity that will allow us to quantify the amount of resources required to execute an algorithm. We then use those quantification methods to introduce the P and NP algorithm classes, and finish this chapter by presenting the concept of NP-completeness as well as some fundamental links between physics and the theory of computation.

Chapter four starts with a succinct introduction to the definitions and main results in the field of discrete classical random walks, followed by concrete examples on the use of classical random walks in algorithm development for solving two versions of the Satisfiability Problem: 2SAT and 3SAT. This chapter finishes with a concise presentation of continuous random walks, a branch of stochastic processes seldom used in computer science but indeed relevant for our further discussion on successful quantum algorithms based on continuous quantum walks.

The second and last part of this lecture, composed of three chapters, provides a solid introduction to the physical, mathematical and computational properties of quantum walks.

Chapter five starts with a comprehensive analysis of a discrete quantum walk on an infinite line, in which we provide relevant definitions plus a study of the mathematical structure that defines the initial conditions, evolution and measurement of a quantum walker whose behavior in space-time is given by a quantum coin and corresponding operators. We then elaborate on the properties and probability distributions of discrete quantum walks on a line with one and two boundaries. This first part of chapter five finishes with the notion of a discrete quantum walk on a graph.

The second part of chapter five starts with an analysis of what is truly quantum about a quantum walk, in which we study some experiments that have shown that some properties of quantum walks are also reproducible by classical systems, and finish this analysis by listing some quantum mechanical properties of quantum walks.

We then proceed to define a continuous quantum walk, followed by a study about the role randomness plays in both continuous and discrete quantum walks. We finish this chapter with a study of a recent, long awaited and very important result: a mathematical connection between discrete and continuous quantum walks.

Chapter six is devoted to the use of both discrete and continuous quantum walks in algorithm development. We start by reviewing some early results on an algorithm based on a discrete quantum walk to find elements in an unordered set; this algorithm employs a hypercube as the geometrical structure on which the quantum walk is performed and uses two definitions of hitting time (i.e. the average time it takes to go from an arbitrary node i to an arbitrary node j) to show that, in both cases, hitting time is of polynomial order.

Chapter six continues with an analysis on a more recent algorithm which shows how to employ a discrete quantum walk to determine whether all elements in a set are distinct or not. We then review a new and refreshing definition of a discrete quantum walk, which consists of a derivation of an evolution operator from a classical stochastic matrix, and finish with a list of relevant papers on more algorithmic applications of quantum walks as well as some results about the impact of decoherence (in this context, decoherence can be understood as performing measurements on the elements of the quantum walk) in the algorithmic performance of a discrete quantum walk.

The second part of chapter six reviews a most celebrated result on quantum

Computation: an algorithm that employs a continuous quantum walk that traverses a family of graphs in polynomial time. This study is followed by a succinct review of an application of a generalized continuous quantum walk for quantum simulation. We finish chapter six by mentioning a very recent and most important result: the computational universality of quantum walks.

Finally, chapter seven provides a summary of our book and proposes some future research directions which, in the opinion of the author, will become increasingly important for both the field of quantum walks and the employment of quantum algorithms in several branches of science.

Acknowledgements

Writing a book is an honor as it allows authors to talk to readers, to share passions and to propose new ideas, regardless of time differences or physical distance. Indeed, it is a matter of justice to acknowledge those who share with a writer the tremendous effort of writing and publishing a book.

I would like to start by acknowledging the loving support of my family for those endless hours that I did not share with them, even during very important festivities like Christmas and New Year's eve. My passion for science has always been supported by the tender patience and encouragement of my mother Amparo, my step-father Bernardo, my sister Samy, and my deceased grandmother Margarita. I warmly thank you all.

I would also like to thank my former DPhil supervisors Professor Keith Burnett and Professor Sougato Bose, who gave me the opportunity to do doctoral research on quantum walks at the University of Oxford. Among those great minds I was privileged to meet during my time in Oxford and who directly contributed to my understanding of quantum walks, I would like to mention Dr Jonathan Ball (who is also the artist who designed the Bloch sphere image used in our chapter on quantum mechanics) and Dr Nikola Paunković.

The idea of writing a book on quantum walks for Morgan & Claypool Publishers came to me from Professor Marco Lanzagorta, whose constant encouragement and friendship are very valuable to me and were crucial components of my enthusiasm during the long nights I spent working on this book. I also want to thank Professor Jeffrey K. Uhlmann who, together with Professor Lanzagorta, are the editors of Morgan & Claypool Synthesis Lectures on Quantum Computing. Furthermore, I want to thank Dr Michael B. Morgan, President of Morgan & Claypool Publishers, for believing in this project and in my capacity to write a book on the novel field of quantum walks.

I have been privileged enough to serve my beloved country, México, working as an assistant professor at Tecnológico de Monterrey Campus Estado de México since 2006. Among the people I have had the honor and the privilege to meet and work for, I would like to mention Professor Pedro Grasa Soler, President of Tecnológico de Monterrey Campus Estado de México, Professor Roberto Rueda Ochoa, President of the Central Zone of Tecnológico de Monterrey, and Professor Ricardo Swain Oropeza, head of the Engineering Division of Tecnológico de Monterrey Campus Estado de México, as their support has been crucial since my first day in campus. Among my colleagues in the departments of computer science and mathematics, and the Quantum Information Processing Group at Tecnológico de Monterrey Campus Estado de México, I thank Professor Edgar Vallejo Clemente, Professor Francisco Delgado Cepeda, Professor Sergio Martínez Casas, Professor José Luis Gómez Muñoz, and Professor Rubén Santiago Acosta for their friendship, wisdom, and desire to do science. I also thank my students, particularly those who took part in the quantum computation lectures I delivered at Tecnológico de Monterrey Campus Estado de México in 2007, for their devotion to knowledge, challenging questions, and inquisitive thoughts. I warmly thank Myrna Romo Serafín for her generosity, support, kind words, and for being a living proof of the everlasting love and unselfish devotion that only outstanding and hard-working mothers like her have for their children.

The final details of this book were written during my stay as visiting Professor at Aspuru-Guzik Group, Harvard University. Among the outstanding people I have met at Harvard and whose friendship will remain forever in my heart, I would like to warmly thank Professor Alán Aspuru-Guzik, Alejandro Perdomo Ortiz, Carolina Perdomo Ortiz, and Roberto Olivares-Amaya.

And, last but never the least, I want to thank the driving forces of Nature, the ultimate responsible entities for the construction of this magnificent world we live in. Every time I think of the tremendously complex system every human body is, and reflect on the fact that I can open my eyes everyday and enjoy the beauty of life and science, I can only feel grateful for the lovely gift life is.

CHAPTER 1

Introduction

Quantum Mechanics and the Theory of Computation are two of the most important intellectual achievements of the 20th century. These two branches of science have not only inspired several generations of scientists and thinkers, they have also had a significant impact on the daily life of mankind, from war to literature (two recent examples of works in the literature inspired by the ideas and history of quantum mechanics are [1, 2]). As a matter of fact, cross-fertilization between physics and computation has been abundant due to the fact that many ideas, concepts, and technological developments from both fields have been used to advance knowledge in each discipline.

One of the most recent joint ventures between physics and the theory of computation is Quantum Computation. Quantum computation can be defined as the scientific field whose purpose is to solve problems with finite time procedures, i.e. algorithms, which exploit the quantum-mechanical properties of those physical systems that are used to implement such algorithms.

Among the theoretical discoveries and promising conjectures that have positioned quantum computation as a key element in modern science, we find (1) the development of novel and powerful methods of computation that may allow us to significantly increase our processing power for solving certain problems [3, 4] and (2) the simulation of complex physical systems that no classical computer would be able, even in principle, to efficiently simulate [5–7]. A detailed summary of scientific and technological applications of quantum computers can be found in [8, 9].

As for the physical realization of quantum computers, several experimental platforms have been developed or customized over the last two decades. Indeed, although it is too early to predict the winning technologies for the implementation of quantum computers, encouraging advances have been made over the last few years in fields such as quantum optics [10–13] and ion traps [14, 15]. Moreover, according to the quantum computation roadmaps produced in the United States of America in 2004 [8] and the European Union in 2007 [9], it is reasonable to expect quantum hardware with enough number of qubits and fault tolerant error correction ready to run quantum simulation and some quantum algorithms by 2012–2017. The reader will find comprehensive lists of physical platforms for quantum computation in [8, 9].

Building good quantum algorithms is a difficult task. First, quantum mechanics is a counterintuitive theory and intuition plays a major role in algorithm design. Second, for a quantum algorithm to be good it is not enough to perform the task it is intended to, but also to do better (i.e. to be more efficient) than any classical algorithm (at least better than those classical algorithms known at the time of developing corresponding quantum algorithm). Examples of successful results in quantum computation can be found in [16–21]. Good introductions to the first quantum algorithms can be found in [22, Ch. 4] and [3, Chs. 4, 5 and 6].

Among those techniques available for the development of quantum algorithms one finds **Quantum Walks**, which is also the main topic of this lecture.

In order to provide a definition of the field of quantum walks, we first introduce the concept of a stochastic algorithm. In the following paragraphs we shall assume that, *in principle*, the problems we intend to solve by using an algorithmic approach are indeed solvable by such a method.

There are several ways to design solutions (i.e. to develop algorithms) in computer science. For example, a powerful method consists of defining a set of rules such that for a given step i in algorithm A, we can always fully determine step $i + 1$, i.e. at any point of the execution of algorithm A we can be fully certain about the next step to be performed, as long as we know the rules of logic used to develop A. Algorithms developed under this methodology are known as **deterministic algorithms** because it is always possible to determine the exact behavior of those algorithms, just by knowing the starting conditions and the set of rules used for algorithm development.

Another method used in algorithm design makes use of chance. In this approach, for a given step s_i of algorithm A, step s_{i+1} cannot be fully determined as there are *several possible next steps*. The actual step $i + 1$ that will be carried out *is chosen* from the set S of possible next steps with the help of a probability distribution. This family of algorithms is known as **stochastic algorithms** and plays the most important role in computer science due to the fact that, in some cases, the most efficient (or least inefficient, depending on the point of view) algorithms known so far for solving certain kinds of problems, are stochastic [23, 24].

A subtle but very important property of stochastic algorithms is the following: assume step s_i of algorithm A is being executed and that there are several possible next steps $s_{i+1} \in S$. Then, all elements of S must have the same probability of being chosen, unless there is a very good reason for giving preference to some computational steps over others (for example, knowledge about the physical properties of a system to be simulated by the stochastic algorithm). The rationale here is that we want chance, i.e. randomness, to determine the evolution of our algorithm. However, if we introduce preferences for some computational steps over others, we are preventing randomness from taking over. Consequently, the generation of random numbers is a fundamental requirement for the execution of stochastic algorithms.

Unfortunately, classical computers are unable to generate truly random numbers [25]. To palliate this disadvantage, computer scientists have developed sophisticated programs for pseudo-random number generation [25] which are good enough for many practical applications of computer science into many branches of science and engineering. However, as true random numbers are still needed, sources of randomness have been produced by using the quantum-mechanical measurement postulate that we will explore in Chapter 2.

Classical random walks, a subset of stochastic processes (that is, processes whose evolution involves chance), have proved to be a very powerful tool for the development of stochastic algorithms [23]. The main idea behind the mechanics of classical random walks is the following: assume we have a particle (walker) that is allowed to move on a lattice. The actual movements of the particle on the lattice, i.e. the evolution of the system, are performed according to a probability distribution. A simple example is the following: suppose that we have a particle on a line, and that the motion of that particle on a line (i.e. moving to the left or to the right) is performed according to the outcomes of tossing a coin (for example, heads → right and tails → left). This process is clearly stochastic and is known as a classical discrete random walk on a line.

Given the importance of classical random walks in algorithm development, there has been an increasing interest in studying quantum counterparts of classical random walks, known as **quantum walks**, in order to develop new quantum algorithms. As we shall see in corresponding chapters, there is already a series of quantum algorithms based on quantum walks that outperform their classical counterparts. Nonetheless, the field of quantum walks is very young and more research is needed to understand how to make full use of this discipline in quantum computation.

There are two main sets of quantum walks: discrete and continuous quantum walks. The main difference between these two sets is the timing used to apply corresponding evolution operators. In the case of discrete quantum walks, the corresponding evolution operator of the system is applied only in discrete time steps, while in the continuous quantum walk case, the evolution operator can be applied at any time.

Our approach in the development of this work has been to study those concepts of quantum mechanics and quantum computation relevant to the computational aspects of quantum walks. Thus, in the history of cross-fertilization between physics and computation, this lecture is meant to be situated as a novel contribution within the field of quantum computation from the perspective of a computer scientist. It has been the intention of this author to write a lecture from which computer scientists with no background in physics may obtain a succinct guide to the concepts of quantum mechanics needed to be initiated in the field of quantum walks.

Although we have assumed that the reader holds a good level of mathematical maturity, particularly in the field of linear algebra, in order to follow certain parts of Chapter 2, we provide a concise list of good introductory texts on that subject in the beginning of Chapter 2.

To the best of author's knowledge, this manuscript is the first book ever published on the topic of quantum walks, and its original approach may allow not only theoretical computer scientists, but also applied computer scientists and engineers to learn the foundations and algorithmic applications of quantum walks.

The following lines provide a summary of the main ideas and contributions of this lecture.

Chapter 2. Quantum Mechanics. This chapter is a concise introduction to the postulates of quantum mechanics (and the mathematical tools required to formulate those postulates) needed to understand the main concepts and techniques of quantum walks, as well as some of the foundational elements of quantum computation. We also provide a brief introduction to entanglement and Bell inequalities.

This chapter has been written with two purposes in mind: (1) to provide the necessary background for our work on quantum walks and (2) to serve as a concise guide for computer scientists who need to grasp those elements of quantum mechanics required to be initiated in the fields of quantum walks and, more generally, quantum computation.

Chapter 3. Theory of Computation. We begin by briefly revisiting the historical roots of the mathematical development of Turing machines, followed by the enunciation of the Church–Turing thesis and the definition of decision problems in the context of computer science. We then proceed to formally define deterministic and nondeterministic models of computation.

We also introduce some formal elements of algorithmic complexity (mainly, measures used to quantify the performance of an arbitrary algorithm), followed by the topic of NP-completeness, one of the central themes in Complexity Theory, together with an example of NP-completeness: the satisfiability (SAT) problem. Finally, we provide a brief review on the links between physics and the theory of computation and give the definitions of Probabilistic and Quantum Turing machines.

Chapter 4. Classical Discrete Random Walks. The goal of this chapter is to provide a short but rigorous introduction to those properties of classical discrete random walks on undirected graphs relevant to algorithm development. We start by offering some basic elements of probability theory (several probability distributions, Markov's inequality and moments of probability distributions), followed by definitions and theorems of Markov chains and stationary probability distributions.

We introduce the definition and main results of classical random walks on a line with three variants: no barriers, one absorbing barrier, and two absorbing barriers. In order to get more general results, we introduce classical random walks on (Cayley) graphs and present two measures used to quantify the performance of classical random walks in algorithm development: hitting time and mixing time.

The last part of this chapter begins with an analysis on the hitting and mixing times of a classical random walk on an unrestricted line. This analysis is, to the best of this author's knowledge, an original contribution to the field of classical random walks, at least in the form that information is presented and the explicit method used to quantify the hitting time of a classical random walk on an unrestricted line.

Basically, we show that the hitting time of a classical discrete random walk on an unrestricted line depends on the region we locate the walker in (we divide the line into two regions: the first one is the area within a distance roughly equal (up to a constant factor) to the standard deviation of the binomial distribution from the starting point of the walk, and the second is the rest of the line). Thus, if we use the hitting time of this random walk to quantify its corresponding mixing time (this is a usual practice in classical random walks), we find that the calculation of the mixing time of a classical random walk on an unrestricted line is not straightforward. This becomes an obstacle for comparing the performance of an unrestricted classical random walk on a line with its quantum counterpart. We will come back to this comparison shortly.

After studying the unrestricted classical random walk on a line, we quantify the hitting and mixing times of classical random walks on a line with two reflecting barriers, and on a circle. We finish this chapter by providing a detailed analysis of the randomized algorithms used to solve two versions of the SAT problem: 2-SAT and 3-SAT, as well as a concise introduction to classical continuous random walks.

Chapter 5. Quantum Walks. In this chapter we offer a comprehensive yet concise introduction to the main concepts, results, and algorithmic applications of discrete quantum walks on a line and on a graph. We first outline the main motivations for doing research in this field, followed by the mathematical description of the components of a quantum walk on a line.

We continue with a detailed analysis of the Hadamard quantum walk on an infinite line, using a method based on the Discrete Time Fourier Transform known as the Schrödinger approach. This analysis includes the enunciation of relevant theorems, as well as the advantages of the Hadamard quantum walk on an infinite line with respect to its closest classical counterpart. In particular, we explore the context in which the properties of the Hadamard quantum walk on an infinite line are compared with classical random walks on an infinite line and with two reflecting barriers. Also, we briefly review another method for studying the Hadamard walk on an infinite line: path counting approach. We then proceed to study a quantum walk on an infinite line with an arbitrary coin operator. In particular, we explain what is meant by stating that the study of the Hadamard quantum walk on an infinite line is enough as for the analysis of arbitrary quantum walks on an infinite line. To finish with our review on quantum walks on a line, we present the main results of quantum walks on a line with one and two absorbing barriers.

We then focus on the properties of quantum walks on graphs. We study quantum walks on a circle, on the hypercube, and some general properties of quantum walks on Cayley graphs. We continue this chapter with an analysis of the connections between classical and quantum walks, followed by a study of subtle but important aspects about the quantumness of quantum walks.

We then proceed to formally define a continuous quantum walk, and focus on issues about the randomness of quantum walks. Finally, we introduce a long-awaited result with respect to connecting the mathematical formalisms of discrete and continuous quantum walks, together with an analysis about whether coins are truly necessary in discrete quantum walks.

Chapter 6. Computer Science and Quantum Walks. We review several links between computer science and quantum walks. We start by introducing how discrete quantum walks can be used to develop quantum algorithms for solving several variants of the search problem, namely the search of M (marked) elements in discrete (and possibly huge) datasets. We then proceed to introduce a novel algorithm based on a mixture of discrete quantum walks and quantum phase estimation for solving combinatorial optimization problems. The second part of this chapter is devoted to analyzing continuous quantum walks. We start by reviewing the most successful quantum algorithm based on a continuous quantum walk known so far, which consists of traversing, in polynomial time, a family of graphs of trees with an exponential number of vertices (the same family of graphs would be traversed only in exponential time by any classical algorithm). We then proceed to briefly review a generalization of a continuous quantum walk, now allowed to perform non-unitary evolution, in order to simulate photosynthetic processes, and we finish by reviewing very recent results about the computational universality of quantum walks.

We finish this introduction with a critical list of articles and books that would provide the reader with further information about the fields we have discussed in this lecture.

Introduction to quantum mechanics and quantum computation: [3, 4, 22, 26–33].
Theory of computation and complexity theory: [34–37].

Classical discrete random walks. Basic concepts of classical random walks can be found in [38–40]. For concepts of classical random walks relevant to algorithm development, the reader may find the following sources useful: [23, 41–43].

Quantum walks. [44] is a good introductory article. For further analysis, we would refer the reader to the research articles and PhD theses cited in Chapters 5 and 6.

CHAPTER 2

Quantum Mechanics

Quantum mechanics is a description of the behavior of matter and light at an atomic scale [27]. Indeed, quantum mechanics plays a fundamental role in the description and understanding of natural phenomena [45].

The history (1900 to *circa* 1930) behind the early experimental and conceptual development of quantum mechanics is a fascinating recollection of scientific experiments and interpretation of experimental results, along with a constant challenge of ideas and assumptions held about Nature for long time [46–48]. Thanks to the works begun by Heisenberg and Schrödinger, and followed by many other physicists like Feynman and Born, quantum mechanics has now a well-developed mathematical structure that provides scientists with a precise theoretical framework with which they can predict the behavior of physical systems. Although there is still debate and controversy about several elements and interpretations of quantum mechanics, using this theory to analyze and predict the behavior of physical systems has proven very fruitful. The birth of Quantum Computation and Quantum Information is a consequence of combining ideas from Quantum Mechanics, Computer Science, and Information Theory.

In this chapter, we introduce those concepts of quantum mechanics needed to understand the main ideas contained in the field of Quantum Walks. In this lecture we have explicitly avoided the topic of interpretations of quantum mechanics, as our interests are focused on the use of quantum mechanics in quantum walks, with the purpose of developing quantum algorithms. Readers interested in interpretative results may be referred to [49, 50] and references mentioned therein.

This chapter begins with some mathematical preliminaries followed by a computer science-oriented presentation of the postulates of quantum mechanics. We then present quantum entanglement and introduce its use as a computational resource, and finish this chapter with a discussion on Bell inequalities.

By definition, learning quantum mechanics in order to develop algorithms is a transdisciplinary task and, consequently, it is a very good idea to learn from different authors and perspectives. Therefore, this chapter is based on [3, 22, 27, 28, 45, 51, 52]. For further learning, we would refer the reader to [53] for a solid yet concise introduction to linear algebra [32, 33] as they are introductions to quantum mechanics and quantum computation for non-physicists [29, 30] for their concise presentations of the postulates of quantum mechanics, and, finally, [54] as it is a useful review of entanglement quantification.

2.1   MATHEMATICAL PRELIMINARIES

In this section we review several concepts of linear algebra required to build the postulates of quantum mechanics. In addition to the references provided at the beginning of this chapter, we would recommend [53] as an excellent introduction to linear algebra. In particular, the reader will find the proofs of all theorems mentioned in this section in [3, 53].

We begin by defining Hilbert spaces, the spaces where mathematical descriptions of quantum physical systems live.

Definition 2.1.1.   Inner-product vector space. *An inner-product vector space* $\mathbb{V}$ *is a complex vector space, equipped with an inner-product* $\langle \cdot | \cdot \rangle : \mathbb{V} \times \mathbb{V} \to \mathbb{C}$, *satisfying the following axioms:* $\forall \, \mathbf{a}, \mathbf{b}, \mathbf{c}, \mathbf{d} \in \mathbb{V}, \alpha, \beta \in \mathbb{C}$
(1) $\langle \mathbf{a} | \mathbf{b} \rangle = \langle \mathbf{b} | \mathbf{a} \rangle^*$,
(2) $\langle \mathbf{a} | \mathbf{a} \rangle \geq 0$ *and* $\langle \mathbf{a} | \mathbf{a} \rangle = 0 \Leftrightarrow \mathbf{a} = \mathbf{0}$,
(3) $\langle \mathbf{a} | \alpha \mathbf{b} + \beta \mathbf{c} \rangle = \alpha \langle \mathbf{a} | \mathbf{b} \rangle + \beta \langle \mathbf{a} | \mathbf{c} \rangle$.[1]
The inner product introduces the **norm** *on* $\mathbb{V}$: $||\mathbf{a}|| = \sqrt{\langle \mathbf{a} | \mathbf{a} \rangle}$.

Definition 2.1.2.   Complete inner-product vector space. *An inner-product vector space* $\mathbb{V}$ *is called* **complete** *if for any sequence* $\{\mathbf{a}_i\}_{i=1}^{\infty}$, $\mathbf{a}_i \in \mathbb{V}$ *with the property* $\lim_{i,j \to \infty} ||\mathbf{a}_i - \mathbf{a}_j|| = 0$, *there is a unique element* $\mathbf{b} \in \mathbb{V}$ *such that* $\lim_{j \to \infty} ||\mathbf{b} - \mathbf{a}_j|| = 0$.

Definition 2.1.3.   Hilbert space. *A* **Hilbert space** $\mathcal{H}$ *is a complete inner-product vector space*[2]. *However, for the purposes of this lecture as well as for understanding the basics of quantum computation, it suffices to define a Hilbert space as an inner-product complex vector space.*

Definition 2.1.4.   Isomorphism among Hilbert spaces. *Two Hilbert spaces* $\mathcal{H}_1$ *and* $\mathcal{H}_2$ *are said to be isomorphic if the underlying vector spaces are isomorphic and their isomorphism preserves the inner product*[3].

---

[1]Rule (3) may also be formulated as $(|u\rangle, \alpha|v\rangle + \beta|w\rangle) = \alpha^*(|u\rangle, |v\rangle) + \beta^*(|u\rangle, |w\rangle)$, where α^* and β^* are the conjugates of α and β, respectively. Nonetheless, it is customary to use Def. 2.1.1 in physics and, particularly, quantum computation.

[2]Complete inner-product spaces were baptized as Hilbert spaces by von Neumann, due to the studies made by Hilbert on linear integral systems. In the following chapter we shall see that Hilbert also played an important role in the birth and development of the Theory of Computation.

[3]In linear algebra, an isomorphism can also be defined as a linear map between two vector spaces that is one-to-one and onto.

Definition 2.1.5. Functional. *Let $\mathbb{V}$ be a vector space over a field F. A **linear functional** is a linear function $f : \mathbb{V} \to F$.*

Lemma 1. *[28]* **Inner-product as linear mapping.** *Let $\mathcal{H}$ be a Hilbert space. Then, for each $\mathbf{a} \in \mathcal{H}$ the function $f_{\mathbf{a}} : \mathcal{H} \to \mathbb{C}$ defined by $f_{\mathbf{a}}(\mathbf{b}) = \langle \mathbf{a} | \mathbf{b} \rangle$ is a linear mapping. Therefore, function $f_{\mathbf{a}}$ is a functional.*

Theorem 1. *[28]. To each continuous linear mapping $f : \mathcal{H} \to \mathbb{C}$ there exists a unique $\phi_f \in \mathcal{H}$ such that $f(\psi) = \langle \phi_f | \psi \rangle$ for any $\psi \in \mathcal{H}$.*

It is possible to prove that the space of all linear functionals of a Hilbert space $\mathcal{H}$ forms again a Hilbert space, the so-called **dual Hilbert space** $\mathcal{H}^*$ over $\mathbb{C}$. Furthermore, Theorem 1 proves that there is a bijection between $\mathcal{H}$ and $\mathcal{H}^*$, therefore $\mathcal{H}$ is isomorphic to $\mathcal{H}^*$. This isomorphism is the basis for the creation of the famous **"bra-ket"** Dirac notation [52].

Definition 2.1.6. Dirac notation. *Let $\mathcal{H}$ be a Hilbert space. A vector $\psi \in \mathcal{H}$ is denoted $|\psi\rangle$ and is referred as a **ket**. The corresponding linear functional is denoted $\langle\psi|$ and is referred to as **bra**. Thus, $\langle\cdot|$ can be seen as an operator that maps each state ϕ into a functional $\langle\phi|$ such that $\langle\phi|(|\psi\rangle) = \langle\phi|\psi\rangle$. We define $|\psi\rangle^\dagger \equiv \langle\psi|$.*

Column and row representation of kets and bras. Let $\mathcal{H}$ be an n-dimensional Hilbert space. Then, $|\psi\rangle \in \mathcal{H}$ can be represented as an n-dimensional column vector, and its corresponding functional $\langle\psi| \in \mathcal{H}^*$ can be seen as an n-dimensional row vector. Therefore, $\langle\phi|\psi\rangle$ is the usual row–column matrix operator that computes the inner product in finite-dimensional vector spaces; $|\psi\rangle \leftrightarrow \langle\psi|$ corresponds to transposition and conjunction.

For example, let $\mathcal{H}^2$ be a two-dimensional Hilbert space, $B_1 = \{|p\rangle, |q\rangle\}$ be a basis of $\mathcal{H}^2$ and $|\psi\rangle \in \mathcal{H}^2$. If α, β are the coefficients of $|\psi\rangle$ with respect to B_1, we can then write $|\psi\rangle = \begin{pmatrix} \alpha \\ \beta \end{pmatrix}$ and $\langle\psi| = (\alpha^*, \beta^*)$.

We now discuss linear operators in Hilbert spaces and their *outer product representation*.

Definition 2.1.7. Linear operator. *Let $\mathcal{H}_1$ and $\mathcal{H}_2$ be Hilbert spaces. Then, a linear operator $\hat{A}$ is a linear function between $\mathcal{H}_1$ and $\mathcal{H}_2$, i.e. $\hat{A} : \mathcal{H}_1 \to \mathcal{H}_2$ such that $\forall |\psi\rangle_i \in \mathcal{H}_1, \alpha_j \in \mathbb{C} \Rightarrow$*

$$\hat{A}\left(\sum_m \alpha_m |\psi\rangle_m\right) = \sum_m \alpha_m \hat{A}(|\psi\rangle_m) = \sum_m \alpha_m |\phi\rangle_m, \quad with \ |\phi\rangle_m \in \mathcal{H}_2.$$

Definition 2.1.8. Outer product representation. *Let $|\psi\rangle, |a\rangle \in \mathcal{H}_1$ and $|\phi\rangle \in \mathcal{H}_2$. Then the **outer product** $|\phi\rangle\langle\psi|$ is the linear operator from $\mathcal{H}_1$ to $\mathcal{H}_2$ defined by*

$$(|\phi\rangle\langle\psi|)(|a\rangle) \equiv |\phi\rangle\langle\psi|a\rangle \equiv \langle\psi|a\rangle|\phi\rangle.$$

Matrix representation of a linear operator. The action of a linear operator $\hat{A}$ is independent of any basis or coordinate system. However, if we choose bases $\{|e\rangle_i\}$ and $\{|f\rangle_i\}$ for $\mathcal{H}_1$ and $\mathcal{H}_2$, respectively, it is possible to give $\hat{A}$ a **matrix representation**. For example, let us define the **Pauli operators** using the matrix representation

$$\sigma_x = \begin{pmatrix} 0 & 1 \\ 1 & 0 \end{pmatrix}; \qquad \sigma_y = \begin{pmatrix} 0 & -i \\ i & 0 \end{pmatrix}; \qquad \sigma_z = \begin{pmatrix} 1 & 0 \\ 0 & -1 \end{pmatrix}. \tag{2.1}$$

Alternatively, we can use the outer product representation

$$\hat{\sigma}_x = |0\rangle\langle 1| + |1\rangle\langle 0|; \qquad \hat{\sigma}_y = i|0\rangle\langle 1| - i|1\rangle\langle 0|; \qquad \hat{\sigma}_z = |0\rangle\langle 0| - |1\rangle\langle 1|, \tag{2.2}$$

where $|0\rangle = \begin{pmatrix} 1 \\ 0 \end{pmatrix}$, $|1\rangle = \begin{pmatrix} 0 \\ 1 \end{pmatrix}$, $\langle 0| = (1,\ 0)$, and $\langle 1| = (0,\ 1)$.

The **Hadamard operator** is another linear operator widely used in quantum walks, its matrix and outer product representations are given by Eqs. (2.3) and (2.4), respectively:

$$H = \frac{1}{\sqrt{2}} \begin{pmatrix} 1 & 1 \\ 1 & -1 \end{pmatrix}, \tag{2.3}$$

$$\hat{H} = \frac{1}{\sqrt{2}} (|0\rangle\langle 0| + |0\rangle\langle 1| + |1\rangle\langle 0| - |1\rangle\langle 1|). \tag{2.4}$$

For example, let us show (in full detail) how the $\hat{\sigma}_x$ operator interacts with an element $|\psi\rangle \in \mathcal{H}^2$, where $\{|0\rangle, |1\rangle\}$ is an orthonormal basis of $\mathcal{H}^2$ and $|\psi\rangle = a|0\rangle + b|1\rangle$:

$$\begin{aligned} \hat{\sigma}_x|\psi\rangle &= (|0\rangle\langle 1| + |1\rangle\langle 0|)(a|0\rangle + b|1\rangle) \\ &= a|0\rangle(\langle 1|0\rangle) + a|1\rangle(\langle 0|0\rangle) + b|0\rangle(\langle 1|1\rangle) + b|1\rangle(\langle 0|1\rangle) \\ &= a(\langle 1|0\rangle)|0\rangle + a(\langle 0|0\rangle)|1\rangle + b(\langle 1|1\rangle)|0\rangle + b(\langle 0|1\rangle)|1\rangle \\ &= (a \times 0)|0\rangle + (a \times 1)|1\rangle + (b \times 1)|0\rangle + (b \times 0)|1\rangle \\ &= 0|0\rangle + a|1\rangle + b|0\rangle + 0|1\rangle \\ &= \mathbf{0} + a|1\rangle + b|0\rangle + \mathbf{0} \\ &= a|1\rangle + b|0\rangle. \end{aligned}$$

The matrix representation of $\hat{\sigma}_x$ and $|\psi\rangle$ can be easily inferred:

$$\sigma_x|\psi\rangle = \begin{pmatrix} 0 & 1 \\ 1 & 0 \end{pmatrix} \begin{pmatrix} a \\ b \end{pmatrix} = \begin{pmatrix} b \\ a \end{pmatrix}.$$

Linear operators given by Eqs. (2.2) and (2.4) are examples of a set of operators widely used in quantum mechanics: **Hermitian** and **unitary** operators.

Lemma 2. *Let $\mathcal{H}$ be a Hilbert space and $\hat{A} : \mathcal{H} \to \mathcal{H}$ a linear operator $\Rightarrow \exists!$ operator $\hat{A}^{\dagger}$, the adjoint of $\hat{A}$, such that $\forall \, |a\rangle, |b\rangle \in \mathcal{H} \Rightarrow \langle a|A|b\rangle = \langle a|A^{\dagger}|b\rangle$. The matrix representation of $\hat{A}^{\dagger}$ is given by $A^{\dagger} = (A^{*})^{T}$, i.e. the conjugate-transpose matrix of A.*

Definition 2.1.9.   Hermitian operator. *Let $\mathcal{H}$ be a finite-dimensional Hilbert space and $\hat{A} : \mathcal{H} \to \mathcal{H}$ a linear operator. If $\hat{A} = \hat{A}^{\dagger}$ then $\hat{A}$ is a* **Hermitian operator**.

Definition 2.1.10.   Positive operator. *Let $\mathcal{H}$ be a Hilbert space and $\hat{A} : \mathcal{H} \to \mathcal{H}$ a linear operator. $\hat{A}$ is a* **positive operator** *if and only if $\forall \, |\psi\rangle \in \mathcal{H} \Rightarrow \langle \psi|\hat{A}|\psi\rangle \geq 0$.*

Definition 2.1.11.   Unitary operator. *Let $\mathcal{H}$ be a Hilbert space and $\hat{U} : \mathcal{H} \to \mathcal{H}$ a linear operator. $\hat{U}$ is a* **unitary operator** *if $\hat{U}\hat{U}^{\dagger} = \hat{I}$, where $\hat{I}$ is the identity operator. Unitary operators are key elements in the formulation of quantum mechanics because they preserve the inner product between vectors: let $|\alpha\rangle = \hat{U}|a\rangle$ and $|\beta\rangle = \hat{U}|b\rangle \Rightarrow \langle \alpha|\beta\rangle = \langle a|\hat{U}^{\dagger}|\hat{U}|b\rangle = \langle a|\hat{I}|b\rangle = \langle a|b\rangle$.*

Unitary and Hermitian operators are examples of normal operators. The mathematical properties of normal operators, particularly the fact that they are diagonalizable, are extremely useful.

Definition 2.1.12.   Normal operator. *Let $\mathcal{H}$ be a Hilbert space and $\hat{A} : \mathcal{H} \to \mathcal{H}$ a linear operator. $\hat{A}$ is normal if $\hat{A}\hat{A}^{\dagger} = \hat{A}^{\dagger}\hat{A}$.*

Theorem 2.   Spectral decomposition. *Any normal operator $\hat{A}$ on a vector space $\mathbb{V}$ is diagonal with respect to some orthonormal basis for $\mathbb{V}$.*

So, a diagonal representation for an operator $\hat{A}$ on a vector space $\mathbb{V}$ is a representation $\hat{A} = \sum_{i} \lambda_{i}|i\rangle\langle i|$, where $\{|i\rangle\}$ is an orthonormal set of eigenvectors for $\hat{A}$ with corresponding eigenvalues λ_{i}. We use this fact to compute operator functions.

Definition 2.1.13.   Operator functions. *Let $f : \mathbb{C} \to \mathbb{C}$ be a function and $\hat{A} = \sum_{i} \lambda_{i}|i\rangle\langle i|$ be a spectral decomposition for a normal operator $\hat{A}$. Then, the operator function $f(\hat{A})$ is defined by*

$$f(\hat{A}) \equiv \sum_{i} f(\lambda_{i})|i\rangle\langle i|.$$

Before we address the topic of creating vector spaces from other vector spaces, we introduce the notions of trace for matrices and linear operators.

Definition 2.1.14.   Trace. *Let $A \in \mathbb{M}_n(F)$ be a matrix of order n with entries (a_{ij}) from field F. The* **trace** *of A is defined as*

$$\text{tr}(A) = \sum_i a_{ii}$$

The trace of a linear operator $\hat{A}$ is defined as the trace of any of its matrix representations [3].

Now we focus on the **tensor product**—a method to build vector spaces from other vector spaces. The tensor product is crucial in representing multiparticle quantum systems.

Definition 2.1.15.   Tensor product. *Let $\mathbb{V}$ and $\mathbb{W}$ be vector spaces (over a field F) of dimension m and n, respectively. Let $\mathbb{X}$ be the tensor product of $\mathbb{V}$ and $\mathbb{W}$, i.e. $\mathbb{X} = \mathbb{V} \otimes \mathbb{W}$. The elements of $\mathbb{X}$ are linear combinations of vectors $|a\rangle \otimes |b\rangle$, where $|a\rangle \in \mathbb{V}$ and $|b\rangle \in \mathbb{W}$. In particular, if $\{|i\rangle\}$ and $\{|j\rangle\}$ are orthonormal bases for $\mathbb{V}$ and $\mathbb{W}$ then $\{|i\rangle \otimes |j\rangle\}$ is a basis[4] for $\mathbb{X}$. Let $\hat{A}, \hat{B}$ be linear operators on $\mathbb{V}$ and $\mathbb{W}$, respectively. Then $\forall\, |a\rangle_1, |a\rangle_2 \in \mathbb{V}$, $|b\rangle_1, |b\rangle_2 \in \mathbb{W}$, and $\alpha \in F \Rightarrow$*
(1) $\alpha(|a\rangle_1 \otimes |b\rangle_1) = (\alpha|a\rangle_1) \otimes |b\rangle_1 = |a\rangle_1 \otimes (\alpha|b\rangle_1)$,
(2) $(|a\rangle_1 + |a\rangle_2) \otimes |b\rangle_1) = |a\rangle_1 \otimes |b\rangle_1 + |a\rangle_2 \otimes |b\rangle_1$,
(3) $|a\rangle_1 \otimes (|b\rangle_1 + |b\rangle_2) = |a\rangle_1 \otimes |b\rangle_1 + |a\rangle_1 \otimes |b\rangle_2$,
(4) $\hat{A} \otimes \hat{B}(|a\rangle_1 \otimes |b\rangle_1) = \hat{A}|a\rangle_1 \otimes \hat{B}|b\rangle_1$.
(5) a generalization of the previous step is straightforward. Let $|a\rangle_i \in \mathbb{V}$, $|b\rangle_i \in \mathbb{W}$, and $\alpha_i \in F \Rightarrow$
$\hat{A} \otimes \hat{B}(\sum_i \alpha_i |a\rangle_i \otimes |b\rangle_i) = \sum_i \alpha_i \hat{A}|a\rangle_i \otimes \hat{B}|b\rangle_i$.

A short-hand notation for $|a\rangle \otimes |b\rangle$ is simply $|ab\rangle$ or $|a, b\rangle$. Also, the tensor product of $|a\rangle$ with itself n times $|a\rangle \otimes |a\rangle \otimes \cdots \otimes |a\rangle$ can also be conveniently written as $|a\rangle^{\otimes n}$.

The **Kronecker product** is a convenient and simple matrix representation of the tensor product. Let $A = (a_{ij})$, $B = (b_{ij})$ be two matrices of orders $m \times n$ and $p \times q$, respectively. Then $A \otimes B$ is given by

$$A \otimes B = \begin{pmatrix} A_{11}B & A_{12}B & \dots & A_{1n}B \\ A_{21}B & A_{22}B & \dots & A_{2n}B \\ \vdots & \vdots & \vdots & \vdots \\ A_{m1}B & A_{m2}B & \dots & A_{mn}B \end{pmatrix}.$$

$A \otimes B$ is of order $mp \times nq$.

Finally, we describe a theorem that will be used in the following section for entanglement quantification. Since we shall use the concept of "pure states" in the following theorem, we ask

---

[4]A concrete example: let $\{|0\rangle, |1\rangle\}$ be an orthonormal basis for a two-dimensional Hilbert space $\mathcal{H}^2$. Then a basis for $\mathcal{H}^2 \otimes \mathcal{H}^2$ is given by $\{|0\rangle \otimes |0\rangle, |0\rangle \otimes |1\rangle, |1\rangle \otimes |0\rangle, |1\rangle \otimes |1\rangle\}$.

the reader to go to **Postulate 1** of the following section in order to review the corresponding definition.

Theorem 3.   Schmidt decomposition. *Suppose $|\psi\rangle$ is a pure state of a composite system $AB \Rightarrow \exists$ orthonormal bases $\{|i_A\rangle\}$ for A and $\{|i_B\rangle\}$ for B such that*

$$|\psi\rangle = \sum_i \lambda_i |i_A\rangle|i_B\rangle,$$

where $\lambda_i \in \mathbb{R}^+ \cup \{0\}$ satisfying $\sum_i \lambda_i^2 = 1$. Numbers λ_i are known as Schmidt coefficients.

2.2   POSTULATES OF QUANTUM MECHANICS

We now provide the postulates of quantum mechanics upon which we build up our work on quantum walks. In quantum mechanics there are two mathematical formalisms to describe a physical quantum system: state vectors and density operators. Both approaches are mathematically equivalent and, consequently, choosing one or the other is a matter of convenient description of the properties of the system to be studied. We formulate Postulates 1–4 in the parlance of state vectors following [3], and additionally define density operators in the context of Postulate 1. Alternative formulations of all postulates in the terminology of density operators can be found in [3, 30, 45].

2.2.1   State Space

The first postulate provides the mathematical framework with which we describe closed (that is, isolated) physical systems.

Postulate 1. To each isolated physical system we associate a Hilbert space $\mathcal{H}$, the **state space** of the system. The physical system is completely described by its **state vector**, which is a unit vector $|\psi\rangle \in \mathcal{H}$. The dimension of $\mathcal{H}$ depends on the specific degrees of freedom of the physical property under consideration. Postulate 1 implies that a linear combination of state vectors is a state vector [45]. This is known as the **superposition principle** and is a quantum-mechanical description of physical systems [45, 52]. In particular, any vector state $|\psi\rangle$ may be described as a superposition of basis states $\{|e_i\rangle\}$ in $\mathcal{H}$, i.e. $|\psi\rangle = \sum_i c_i |e_i\rangle$, $c_i \in \mathbb{C}$.

An alternative description of quantum states is given by the **density operator** (also called **density matrix**). The density operator is positive Hermitian and has trace equal to 1. A quantum system whose state $|\psi\rangle$ is exactly known is said to be in a **pure state**. The density operator of a pure state is given by $\hat{\rho} = |\psi\rangle\langle\psi|$. A density operator also describes **mixed quantum states**. A mixed state may be obtained from a source randomly producing pure states. For example, suppose that a quantum system has a quantum state picked up from a set of possible quantum states $\{|\psi\rangle_i\}$ according to a probability distribution $\{p_i\}$. Then its density operator is given by

$$\hat{\rho} = \sum_i p_i |\psi\rangle_i \langle\psi|. \tag{2.5}$$

Density operators do not uniquely represent a probability distribution over pure states, as it is possible to have two different quantum state ensembles giving rise to the same density operator.

The Qubit

In classical computation, information is stored and manipulated in the form of bits. The mathematical structure of a classical bit is rather simple. It suffices to define two "logical" values, traditionally labeled as $\{0, 1\}$, and to relate these values to two different outcomes of a classical measurement (for example, in TTL transistor technology, "0" is a label given to a voltage measurement between 0 V and 0.5 V, while "1" is the label attached to a voltage measurement between 4.5 V and 5 V). So, a classical bit "lives" in a scalar space.

In quantum computation, information is stored, manipulated, and measured in the form of qubits. A qubit is a physical entity described by the laws of quantum mechanics. Simple examples of qubits include two orthogonal polarizations of a photon (e.g. horizontal and vertical), the alignment of a (spin-1/2) nuclear spin in a magnetic field or two states of an electron orbiting an atom. A qubit may be mathematically represented as a unit vector in a two-dimensional Hilbert $|\psi\rangle \in \mathcal{H}^2$. A qubit $|\psi\rangle$ may be written in general form as

$$|\psi\rangle = \alpha|p\rangle + \beta|q\rangle, \tag{2.6}$$

where $\alpha, \beta \in \mathbb{C}$, $|\alpha|^2 + |\beta|^2 = 1$, and $\{|p\rangle, |q\rangle\}$ is an arbitrary basis spanning $\mathcal{H}^2$. The choice of $\{|p\rangle, |q\rangle\}$ is often $\{|0\rangle, |1\rangle\}$, the so-called computational basis states which form an orthonormal basis for $\mathcal{H}^2$. In general, $|\psi\rangle$ is a coherent superposition of the basis states $|p\rangle$ and $|q\rangle$ and can be prepared in an infinite number of ways simply by varying the values of the complex coefficients α and β subject to the normalization constraint.

We can rewrite Eq. (2.6) as

$$|\psi\rangle = e^{i\gamma}\left(\cos\frac{\theta}{2}|0\rangle + e^{i\varphi}\sin\frac{\theta}{2}|1\rangle\right), \tag{2.7}$$

where $\gamma, \theta,$ and $\varphi \in \mathbb{R}$. Since $e^{i\gamma}$ has no observable effects [3] (i.e. measurement outcomes are invariant to any value of γ) we can ignore it. Thus

$$|\psi\rangle = \cos\frac{\theta}{2}|0\rangle + e^{i\varphi}\sin\frac{\theta}{2}|1\rangle. \tag{2.8}$$

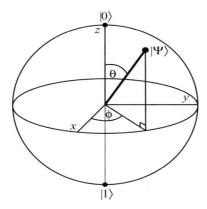

FIGURE 2.1: Bloch sphere representation of a qubit $|\psi\rangle = \cos\frac{\theta}{2}|0\rangle + e^{i\phi}\sin\frac{\theta}{2}|1\rangle$.

The numbers θ and φ define a point on the unit three-dimensional sphere known as **Bloch sphere** (Fig. 2.1).

It is a good idea to use a vector representation in problems where we know with certainty the initial state of the qubit. An example of this statement is to prepare a qubit in the state $|\Psi\rangle = \frac{|0\rangle+|1\rangle}{\sqrt{2}}$, that is, an equally weighted superposition of the canonical basis $\{|0\rangle, |1\rangle\}$.

However, let us consider a different scenario in which a qubit $|\Psi\rangle$ is initially prepared in one of the following quantum states: $\{|\psi\rangle_1, |\psi\rangle_2, |\psi\rangle_3, \ldots, |\psi\rangle_n\}$ where each of the states is selected with probability $\frac{1}{n}$. We do not know what state was chosen to prepare $|\Psi\rangle$, but we do know that only preparations $|\psi\rangle_i$, $i \in \{1, 2, \ldots, n\}$, are allowed. In this case, a convenient representation for $|\Psi\rangle$ is the associated density operator $\hat{\rho}_\Psi = \frac{1}{n}\sum_{k=1}^{n}|\psi\rangle_k\langle\psi|$.

2.2.2 Evolution of a Closed Quantum System

Postulate 2 (Unitary operator version). The evolution of a closed quantum system with state vector $|\Psi\rangle$ is described by a unitary transformation $\hat{U}$ (Def. 2.1.11). The state of a system at time t_2 according to its state at time t_1 is given by

$$|\Psi(t_2)\rangle = \hat{U}|\Psi(t_1)\rangle. \tag{2.9}$$

Postulate 2 only describes the mathematical properties that an evolution operator must have. The specific evolution operator required to describe the behavior of a particular quantum system depends on the system itself. In the case of single qubits, any unitary operator can be realized in physical systems [3]. Postulate 2 can also be stated with the famous **Schrödinger equation**.

Postulate 2 (Hermitian operator version). The evolution of a closed quantum system is described by the Schrödinger equation

$$i\hbar\frac{d|\psi\rangle}{dt} = \hat{\mathbf{H}}|\psi\rangle, \tag{2.10}$$

where $\hbar$ is Planck's constant and $\hat{\mathbf{H}}$ is a fixed Hermitian operator (Eq. (2.1.9)) known as the *Hamiltonian* of the closed system (note that in spite of similar notation, $\hat{\mathbf{H}}$ and $\hat{H}$ represent two different things, being the former the Hamiltonian of Postulate 2 and the latter the Hadamard operator). The Hamiltonian of particular physical systems must be determined and calculated for each case. In general, figuring out the Hamiltonian of a particular physical system is a difficult task.

In this lecture we make extensive use of the Hadamard operator (Eq. (2.4)) as evolution operator, among others. The effect of the Hadamard operator as evolution operator is exemplified in the following two equations:

$$\hat{H}|0\rangle = \frac{1}{\sqrt{2}}[|0\rangle\langle 0| + |0\rangle\langle 1| + |1\rangle\langle 0| - |1\rangle\langle 1|]|0\rangle = \frac{1}{\sqrt{2}}(|0\rangle + |1\rangle),$$

$$\hat{H}|1\rangle = \frac{1}{\sqrt{2}}[|0\rangle\langle 0| + |0\rangle\langle 1| + |1\rangle\langle 0| - |1\rangle\langle 1|]|1\rangle = \frac{1}{\sqrt{2}}(|0\rangle - |1\rangle).$$

2.2.3   Quantum Measurements

In quantum mechanics, measurement is a non-trivial and highly counter-intuitive process for several reasons: first, because measurement outcomes are inherently probabilistic, i.e. regardless of the carefulness in the preparation of a measurement procedure, the possible outcomes of such measurement will be distributed according to a certain probability distribution. Second, once a measurement has been performed, a quantum system in unavoidably altered due to the interaction with the measurement apparatus. Consequently, for an arbitrary quantum system, pre-measurement and post-measurement quantum states are different in general. Third, in order to perform a measurement it is needed to define a set of measurement operators. This set of operators must fulfil a number of rules that allows one to compute the actual probability distribution as well as post-measurement quantum states.

Postulate 3. Quantum measurements are described by a set of measurement operators $\{\hat{O}_n\}$, index n labels the different measurement outcomes, which act on the state space of the system being measured. Measurement outcomes correspond to the values of *observables*, such as position, energy, and momentum, which are Hermitian operators (Def. 2.1.9) corresponding to physically measurable quantities.

A projective measurement is described by an *observable* $\hat{M}$, a Hermitian operator (Def. 2.1.9) defined in the state space of the quantum system. By using the spectral decomposition

Theorem 2, it is possible to write $\hat{M}$ as follows:

$$\hat{M} = \sum_i r_i \hat{P}_{r_i}, \tag{2.11}$$

where $\hat{P}_{r_i}$ is the projector operator for the eigenspace $E(r_i)$ defined by the eigenvalue r_i. Measurement outcomes correspond to the eigenvalues r_i of observable $\hat{M}$.

Let $|\psi\rangle$ be the state of the quantum system immediately before the measurement. Then, the probability that result r_i occurs is given by

$$p(r_i) = \langle \psi | \hat{P}_{r_i} | \psi \rangle, \tag{2.12}$$

and the post-measurement quantum state $|\psi\rangle_{\mathrm{pm}}$ associated with outcome r_i is given by

$$|\psi\rangle_{\mathrm{pm}} = \frac{\hat{P}_{r_i} |\psi\rangle}{\sqrt{p(r_i)}}. \tag{2.13}$$

Let us work out a simple example. Assume we have a photon with associated polarization orientations "horizontal" and "vertical." The horizontal polarization direction is denoted by $|0\rangle$ and the vertical polarization direction is denoted by $|1\rangle$. By Postulate 1 (Eq. (2.6)), an arbitrary initial state for our photon can be described by the quantum state

$$|\psi\rangle = \alpha |0\rangle + \beta |1\rangle, \tag{2.14}$$

where α and β are complex numbers constrained by the normalization condition $|\alpha|^2 + |\beta|^2 = 1$, and $\{|0\rangle, |1\rangle\}$ is the computational basis spanning $\mathcal{H}^2$.

In other words, the polarization of an arbitrary photon is neither only vertical or only horizontal, but a linear combination of both polarizations. One way to look at Eq. (2.14) is to think of the corresponding photon as having both vertical *and* horizontal polarization simultaneously. If we think of photon polarization as a degree of freedom for storing information, we may think of such a photon as a qubit with both "0" and "1" values simultaneously stored in its polarization.

Now we construct two measurement operators $\hat{P}_{a_0} = |0\rangle\langle 0|$ and $\hat{P}_{a_1} = |1\rangle\langle 1|$ with corresponding measurement outcomes r_0, r_1. Then, the full *observable* used for measurement in this experiment is $\hat{M} = r_0 |0\rangle\langle 0| + r_1 |1\rangle\langle 1|$. According to Postulate 3, we can say the following:

(1) There are only two possible measurement outcomes for observable $\hat{M}$: r_0 and r_1.

(2) According to Eq. (2.12), the probability of obtaining the measurement outcome r_0 is

$$p(r_0) = \langle \psi | \hat{P}_{r_0} | \psi \rangle = (\langle 1 | \beta^* + \langle 0 | \alpha^*) \hat{P}_{r_0} (\alpha |0\rangle + \beta |1\rangle) = |\alpha|^2.$$

(3) If, after measuring the polarization degree of freedom of our photon we find that measurement outcome is indeed r_0 then its corresponding post-measurement quantum state

$|\psi\rangle_{\text{pm}}^{r_0}$ is given by Eq. (2.13), namely,

$$\frac{\hat{P}_{a_0}|\psi\rangle}{\sqrt{p(a_0)}} = \frac{|0\rangle\langle 0|(\alpha|0\rangle + \beta|1\rangle)}{\sqrt{|\alpha|^2}} = |0\rangle.$$

(4) In the same vein and according to Eq. (2.12), the probability of obtaining measurement outcome r_1 is

$$p(a_1) = \langle\psi|\hat{P}_{r_1}|\psi\rangle = (\langle 1|\beta^* + \langle 0|\alpha^*)\hat{P}_{r_1}(\alpha|0\rangle + \beta|1\rangle) = |\beta|^2.$$

(5) If measurement outcome is indeed r_1 then its corresponding post-measurement quantum state $|\psi\rangle_{\text{pm}}^{r_1}$ is given by Eq. (2.13), that is

$$\frac{\hat{P}_{r_1}|\psi\rangle}{\sqrt{p(r_1)}} = \frac{|1\rangle\langle 1|(\alpha|0\rangle + \beta|1\rangle)}{\sqrt{|\beta|^2}} = |1\rangle.$$

Before introducing the last postulate of quantum mechanics, which states how multipartite quantum systems must be mathematically represented, let us present a highly non-trivial consequence of Postulates 1 and 2: the no-cloning theorem.

The No-cloning Theorem

A trivial assumption in algorithm development within the field of classical computer science is the fact that we can copy as many bits as required. This assumption, based on the properties of classical physical systems, is so deeply rooted in us that the technical procedures used to copy electrical signals in (say) computer memories are usually not studied in computer science textbooks but in electrical and electronics engineering textbooks only.

However, this capacity of making exact copies of arbitrary information contained in physical systems is not present in the domain of quantum computation. In other words, in general it is impossible to make exact copies of the value contained in one arbitrary qubit into another qubit. This counter-intuitive (and somewhat discouraging, at least apparently) result was proved in [55, 56]. Now let us go through the details of this proof (*reductio ad absurdum*).

Let A, B be quantum physical systems described by qubits $|\psi\rangle_A$ and $|\phi\rangle_B$, respectively. The qubit $|\psi\rangle_A$, a pure but unknown quantum state, contains the (arbitrary) information we want to copy, while $|\phi\rangle_B$ is the qubit in which we would like to copy the information contained in $|\psi\rangle_A$.

The capacity of copying the information contained in $|\psi\rangle$ into $|\phi\rangle$ by quantum evolution can be written as

$$\hat{U}(|\psi\rangle_A|\phi\rangle_B) = |\psi\rangle_A|\psi\rangle_B. \tag{2.15}$$

The purpose of a hypothetical copying quantum circuit is to allow us to copy arbitrary information. Thus the circuit should work for two different and unknown descriptions of physical system A, denoted by $|\psi\rangle_A$ and $|\omega\rangle_A$:

$$\hat{U}(|\psi\rangle_A|\phi\rangle_B) = |\psi\rangle_A|\psi\rangle_B \tag{2.16}$$

and

$$\hat{U}(|\omega\rangle_A|\phi\rangle_B) = |\omega\rangle_A|\omega\rangle_B. \tag{2.17}$$

However, if we compute the inner product of Eqs. (2.16) and (2.17), we find that

$$_A\langle\phi|_B\langle\psi|\hat{U}^\dagger|\hat{U}|\omega\rangle|_A|\phi\rangle_B = \langle\psi|\omega\rangle = ((\langle\psi|\omega\rangle))^2. \tag{2.18}$$

Note that $\langle\psi|\omega\rangle = ((\langle\psi|\omega\rangle))^2$ implies either $\langle\psi|\omega\rangle = 1$ (i.e. $|\psi\rangle = |\omega\rangle$) or $\langle\psi|\omega\rangle = 0$ (i.e. $|\psi\rangle \perp |\omega\rangle$). Therefore, qubit cloning works only when the two states are either identical (which contradicts our hypothesis) or orthogonal [31]. Orthogonal qubits produce deterministic measurement outcomes for a given set of measurement projector operators $\{\hat{P}_i\}$; in other words, orthogonal qubits contain information that behaves exactly the same way as classical bits, since labels "0" and "1" represent mutually exclusive measurement outcomes of a classical physical system (for example, voltage values in a transistor).

Before finishing this subsection, we would like to underline that the impossibility of pure quantum state cloning *does not mean* that *imperfect* copying is not possible. Indeed, imperfect pure and mixed quantum state cloning is an active area of research (see [57] for a review).

2.2.4   Composite Quantum Systems

We now focus on the mathematical description of a composite quantum system, i.e. a system made up of several different physical systems.

Postulate 4. The state space of a composite quantum system is the tensor product of the component system state spaces.

- If we have n quantum systems expressed as *state vectors*, labeled $|\psi\rangle_1, |\psi\rangle_2, \ldots, |\psi\rangle_n$, then the joint state of the total system is given by $|\psi\rangle_T = |\psi\rangle_1 \otimes |\psi\rangle_2 \otimes \cdots \otimes |\psi\rangle_n$.
- Similarly, if we have n quantum systems expressed as *density operators* $\rho_1, \rho_2, \ldots, \rho_n$ then the joint state of the total system is given by $\rho_T = \rho_1 \otimes \rho_2 \otimes \cdots \otimes \rho_n$ (in the absence of any knowledge of correlations).

As an advance of the operations we shall perform on the following chapters, let us show the details of applying an evolution operator to a composite quantum system. Let $\hat{H}^{\otimes 2}$ be the tensor product of the Hadamard operator (Eq. (2.4)) with itself and let $|\psi\rangle = |00\rangle$. Then

$$
\begin{aligned}
\hat{H}^{\otimes 2} |\psi\rangle = \frac{1}{2}(&|00\rangle\langle00| + |01\rangle\langle00| + |10\rangle\langle00| + |11\rangle\langle00| + |00\rangle\langle01| - |01\rangle\langle01| \\
&+ |10\rangle\langle01| - |11\rangle\langle01| + |00\rangle\langle10| + |01\rangle\langle10| - |10\rangle\langle10| - |11\rangle\langle10| + |00\rangle\langle11| \\
&- |01\rangle\langle11| - |10\rangle\langle11| + |11\rangle\langle11|)|00\rangle = \frac{1}{2}(|00\rangle + |01\rangle + |10\rangle + |11\rangle).
\end{aligned}
$$

$$(2.19)$$

2.3 ENTANGLEMENT

Entanglement is a unique type of correlation shared between components of a quantum system. Entangled quantum systems are sometimes best used collectively, that is, sometimes an optimal use of entangled quantum systems for information storage and retrieval includes manipulating and measuring those systems as a whole, rather than on an individual basis. Quantum entanglement and the principle of superposition are two of the main concepts behind the power of quantum computation and quantum information theory.

The concept of correlation is deeply rooted in every branch of science. A typical and simple example is the following experiment: let us suppose we have two balls, one white and one black, as well as two boxes. If we randomly put a ball in each box and then close both boxes, we need to perform only one experiment, that is, to open one box, in order *to know which of the balls is in each box*. In other words, by means of one measurement, namely opening one box and seeing which ball was stored in it, we obtain two pieces of information, namely the color of the ball stored in each box.

The former experiment is an example of classical correlation. Quantum entanglement is also a kind of correlation, but one that is detected only in quantum phenomena. A good example of the difference between classical and quantum correlations would be correlations in canonically conjugate observables, such as position and momentum.

Consider the following 2-particle state:

$$
|\Psi^-\rangle = \frac{|01\rangle - |10\rangle}{\sqrt{2}}.
$$

$$(2.20)$$

Clearly, $|\Psi_-\rangle$ lives in a four-dimensional Hilbert space. It can be seen, after some calculations, that it is impossible to find quantum states $|a\rangle, |b\rangle \in \mathcal{H}^2$ such that $|a\rangle \otimes |b\rangle = |\Psi_-\rangle$, that is, $|\Psi_-\rangle$ is not a product state of $|a\rangle$ and $|b\rangle$. This is indeed a criterion to determine whether a quantum state is entangled or not, whether it is possible to express such a composite quantum state as a simple tensor product of quantum subsystems. Another example is the tripartite entangled GHZ state

$$
|\text{GHZ}\rangle = \frac{|000\rangle + |111\rangle}{\sqrt{2}}.
$$

$$(2.21)$$

Again, it is not possible to find three quantum states $|a\rangle, |b\rangle, |c\rangle \in \mathcal{H}^2$ such that $|a\rangle \otimes |b\rangle \otimes |c\rangle = |\text{GHZ}\rangle$. Entanglement definition and quantification is an open research area. Currently it is known how to identify and quantify entanglement for two particles, but for three or more particles the situation is far less straightforward and remains an active area of research.

2.3.1   Measure of Entanglement

The von Neumann entropy of the reduced density operator is a mathematical tool used to quantify the degree of entanglement of the quantum systems. With the purpose of formally introducing this measure, let us first present the concept of reduced density operator.

Reduced Density Operator

Let us suppose we have a density operator describing a composite quantum system C and we are interested in studying the properties of one subsystem of C (such a situation would happen, for example, if after creating a bipartite quantum system we had access to only one particle). The description of such a subsystem is provided by the reduced density operator, defined as follows.

Definition 2.3.1.   *Let A, B be two physical systems whose state is described by a density operator ρ^{AB}. The reduced density operator for system A is defined as*

$$\rho^A \equiv tr_B(\rho^{AB}),$$

where tr_B is the partial trace over system B. The partial trace is given by

$$tr_B(|\alpha_1\rangle\langle\alpha_2| \otimes |\beta_1\rangle\langle\beta_2|) \equiv |\alpha_1\rangle\langle\alpha_2| tr(|\beta_1\rangle\langle\beta_2|) \equiv |\alpha_1\rangle\langle\alpha_2|\langle\beta_2|\beta_1\rangle.$$

von Neumann Entropy of the Reduced Density Operator

For a pure quantum state $|\psi\rangle$ of a composite system AB with $\dim(A) = d_A$ and $\dim(B) = d_B$, let $|\psi\rangle = \sum_{i=1}^{d} \alpha_i |i_A\rangle |i_B\rangle$ ($d = \min(d_A, d_B)$, $\alpha_i \geq 0$, and $\sum_{i=1}^{d} \alpha_i^2 = 1$) be its Schmidt decomposition. Also, let $\rho_A = tr_B(|\psi\rangle\langle\psi|)$ and $\rho_B = tr_A(|\psi\rangle\langle\psi|)$ be the reduced density operators of systems A and B, respectively. *The entropy of entanglement $E(|\psi\rangle)$ is the von Neumann entropy of the reduced density operator* [3, 54, 58]:

$$E(|\psi\rangle) = S(\rho_A) = S(\rho_B) = -\sum_{i=1}^{d} \alpha_i^2 \log_2(\alpha_i^2). \tag{2.22}$$

E is a monotonically increasing function of the entanglement present in the system AB. A non-entangled state has $E = 0$. States $|\psi\rangle \in \mathcal{H}^d$ for which $E(\psi) = \log_2 d$ are called *maximally entangled states* in d dimensions. In particular, note that for the quantum states (known as *Bell states*) $|\Phi^+\rangle = \frac{1}{\sqrt{2}}(|00\rangle + |11\rangle)$, $|\Phi^-\rangle = \frac{1}{\sqrt{2}}(|00\rangle - |11\rangle)$, $|\Psi^+\rangle = \frac{1}{\sqrt{2}}(|01\rangle + |10\rangle)$, and $|\Psi^-\rangle = \frac{1}{\sqrt{2}}(|01\rangle - |10\rangle)$, $E(|\Phi^+\rangle) = E(|\Phi^-\rangle) = E(|\Psi^+\rangle) = E(|\Psi^-\rangle) = 1$, i.e. these states are maximally entangled.

2.3.2   Bell Inequalities

Bell inequalities are a powerful tool for entanglement detection. Thus, for the sake of providing a brief introduction to Bell inequalities for computer scientists as well as for the potential use of these ideas in physical realizations of quantum computers, we discuss some of the main concepts behind Bell inequalities.

The counter-intuitive properties of quantum mechanics have always been a source of controversy. In their seminal paper [59], Einstein, Podolsky, and Rosen (EPR) proposed a thought experiment with which they tried to show that quantum mechanics was an incomplete theory of Nature. The thought experiment proposed in [59] was developed under the following lines of thought.

1. **Assumption of Realism**. Physical properties have definite values which exist independent of observation.

2. **Assumption of Locality**. The description of a system's state depends only in itself and its immediate surroundings. Therefore, for sufficiently separated physical systems, measurements performed on one of them cannot have any influence on the others.

These two assumptions together are known as **local realism**. According to [59], quantum mechanics was an incomplete theory under a local realistic description of Nature.

The discussion about the controversial properties of quantum mechanics shown in [59] was considered to be just philosophical for long time. However, in 1964, Bell published [51], in which he derived an inequality (involving correlated measurement results) that would have to be obeyed by any system behaving under the rules of local realism. Furthermore, it was also shown that for some entangled systems the inequality would be violated. Naturally, testing whether Nature was in fact local-realistic became an appealing idea.

A number of experiments [22, p. 12] have shown strong evidence that the inequality proposed in [51] is not obeyed by Nature[5]. Furthermore, the quantum-mechanical prediction was confirmed. The violation of Bell inequalities implies that at least one of the assumptions of

---

[5]It must be noted that there is still controversy on the invalidity of local realism, at least in written evidence. For example, it was written on an essay by Bouwmeester and Zeilinger [22, p. 12] that "Even though a number of experiments have now confirmed the quantum predictions, from a strictly point of view the problem is not closed yet as some loopholes in the existing experiments still make it logically possible, at least in principle, to uphold a local realist world view."

local realism is in conflict with quantum mechanics. Although this is usually viewed as evidence for non-locality, there are some other possible explanations [22, 50].

In addition to its relevance to the foundations of physics and as stated in the beginning of this section, Bell inequalities can be used as a resource to detect entanglement in certain cases (for example, see [60, 61]).

CHAPTER 3

Theory of Computation

The purpose of this chapter is to concisely introduce the *Theory of Computation*, in order to provide the necessary background and to motivate our further discussion on the importance of classical random walks and quantum walks in computer science.

We begin by providing an overview of the theory of computation and deliver a succinct historical background of those ideas that led to its creation and development. We then introduce a formal definition of a model of computation that has played the most important role in the development of Computer Science: Turing Machines. The previous concepts are followed by key definitions and theorems from Complexity Theory and the definitions of **P, NP**, and **NP-complete** problem categories. In the last section of this chapter in which we discuss some of the liasons between physics and computation, we also introduce another model of computation: Quantum Turing Machines. This chapter is based on [34, 35, 37, 62, 63].

3.1 WHAT IS THE THEORY OF COMPUTATION?

The Theory of Computation is a scientific field devoted to understanding the fundamental capabilities and limitations of computers, and is divided into three areas: (1) *Automata Theory*, or the study of different models of computation, (2) *Computability Theory*, which focuses on determining which problems can be solved by computers and which cannot, and (3) *Complexity Theory*, devoted to studying what makes some problems computationally hard and other easy.

The development of the theory of computation was driven in great part by several challenges posed by D. Hilbert and other mathematicians on the foundations of mathematics at the beginning of the 20th century [64]. Turing and other scientists, while working on the ideas required to formalize the idea of computation, answered some of the questions posed by Hilbert et al.

3.2 THE BIRTH OF THE THEORY OF COMPUTATION: ALAN TURING AND HIS MACHINES

In 1936, Alan Turing published the most influential paper [65] in which he pioneered the theory of computation, introducing the famous abstract computing machines now known as

Turing Machines. In [65], Turing explained the fundamental principle of the modern computer, the idea of controlling the machine's operation by means of a program of coded instructions stored in the computer's memory, i.e. Turing showed that it was possible to build a *Universal Turing Machine*, that is, a Turing machine capable of simulating any other Turing machine (in nowadays jargon, the Universal Turing Machine would be the actual digital computer and the simulated Turing machine the program that has been encoded in the digital computer's memory).

Among Turing's key contributions from [65], we find: (1) a definition of procedure, which evolved into the modern definition of algorithm, (2) the Church–Turing thesis: every function that would be naturally regarded as computable can be computed by the Universal Turing Machine, and (3) the definition of a Turing machine.

Now we shall briefly review the definitions of deterministic and nondeterministic computation, together with the definitions of deterministic and nondeterministic Turing machines.

3.3   DETERMINISTIC AND NONDETERMINISTIC COMPUTATION

When every step of a computation follows in a unique way from the preceding step, we are doing deterministic computation. In the nondeterministic computation, several choices may exist for the next state at any point.

How does a nondeterministic machine (NDM) compute? Suppose that we are running an NDM on an input string and reach a state with multiple states to proceed. At this point, *the machine splits into multiple copies of itself and follows all the possibilities in parallel.* Each copy of the machine takes one of the possible ways to proceed and continues as before. In the case of subsequent choices, the machine splits again. If the next input symbol does not appear on any of the arrows exiting the state occupied by a copy of the machine, that copy of the machine dies. Finally, if any one of these copies of the machine is in an accept state at the end of the input, the NDM accepts the input string.

Another way to think of a nondeterministic computation is as a tree of possibilities. The root of the tree corresponds to the start of the computation. Every branching point in the tree corresponds to a point in the computation at which the machine has multiple choices. The machine accepts if at least one of the computation branches ends in an accept state. A graphical illustration of a nondeterministic computation is given in Fig. 3.1. Note that nondeterminism is a generalization of determinism.

Under the laws of classical physics, nondeterminism is not a fully realistic model of computation as it assumes the capability of producing several instances of a machine to run in parallel: it would be like suddenly producing as many computers as instances for each computation step. The importance of nondeterministic computation is based on having a

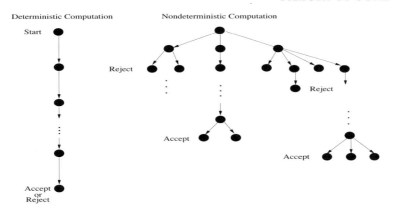

FIGURE 3.1: In a deterministic computation, every single step is fully determined by the previous step. In contrast, in nondeterministic computation, a step may be followed by m new steps or, equivalently, the NDM makes m copies of itself, one for each possibility.

powerful computational model in which we can find out whether a problem can be solved *in principle*, regardless of the amount of resources employed for that task.

However, there has been significant efforts toward finding physical methods for implementing nondeterministic computation. For example, nondeterminism may be viewed as a kind of pseudo-parallel computation wherein multiple independent processes can be running concurrently (an NDM splitting to follow several choices may be thought of as a process forking into several children, each proceeding separately).

Another method is to randomly choose a branch from a nondeterministic computation tree. Although this stochastic approach is not entirely equivalent to following all possible branches of the same tree, this method has proven very fruitful to attack several key problems in computer science, known as NP-complete problems. We shall briefly review the ideas behind NP-completeness in the following section.

A third approach toward implementing nondeterministic computation is to incorporate quantum mechanics in algorithm development, for the following reasons: (1) quantum superposition can be used to compute several instances of the same problem (for example, computing the values of a function for many elements in its domain). This is known as quantum parallelism [66]. (2) Furthermore, the probabilistic nature of quantum measurement may be used to randomly choose branches from a nondeterministic computation tree.

Now let us briefly remind the reader of the definitions of the deterministic and nondeterministic Turing machines. These definitions will be very useful for providing a mathematical description of a quantum Turing machine.

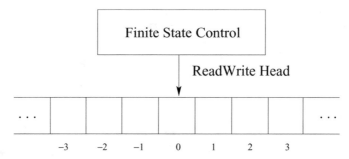

FIGURE 3.2: The "hardware" elements of a Deterministic Turing Machine (DTM) are a limitless memory-type (the tape is divided into squares or cells), and a scanner which consists of read–write head *plus* a finite state control system. The scanner has two purposes: to read and write information on the cells of the tape as well as to control the state of the DTM.

A Deterministic Turing Machine (DTM), pictured schematically in Fig. 3.2, is an accurate model of a real general purpose computer. A mathematical definition of a DTM follows:

Definition 3.3.1.   Deterministic Turing Machine. *A Deterministic Turing Machine (DTM) is a 7-tuple* $M = (Q, \Sigma, \Gamma, \delta, q_0, q_{accept}, q_{reject})$, *where* Q, Σ, Γ *are all finite sets, an alphabet is by definition a finite set, and*

1. Q is the set of states,

2. Σ is the input alphabet not containing the blank symbol $\sqcup$,

3. Γ is the tape alphabet, where $\sqcup \in \Gamma$ and $\Sigma \subset \Gamma$,

4. $\delta : Q \times \Gamma \to Q \times \Gamma \times \{L, R\}$ is the transition function,

5. $q_0 \in Q$ is the start state,

6. $q_{accept} \in Q$ is the accept state, and

7. $q_{reject} \in Q$ is the reject state, where $q_{accept} \neq q_{reject}$.

A problem is decidable by a DTM if there is an algorithm that halts on all inputs after a finite number of steps. A language is decidable if it is associated with a decidable problem. A counterexample of a decidable problem, i.e. a problem for which no Turing machine can be built in order to solve it, is the Halting problem [65].

A DTM is a computational model in which we can implement any deterministic algorithm designed up to date. We now expand the definition of a Turing machine in order to use it for nondeterministic computation.

Definition 3.3.2. Nondeterministic Turing Machine. *A Nondeterministic Turing Machine is a 7-tuple* $M_N = (Q, \Sigma, \Gamma, \Delta, q_0, q_{accept}, q_{reject})$, *where* Q, Σ, Γ *are all finite sets,* $\mathcal{P}(Q \times \Gamma \times \{L, R\})$ *is the power set of* $Q \times \Gamma \times \{L, R\}$, *and*

1. Q *is the set of states,*
2. Σ *is the input alphabet not containing the blank symbol* $\sqcup$,
3. Γ *is the tape alphabet, where* $\sqcup \in \Gamma$ *and* $\Sigma \subset \Gamma$,
4. $\Delta : Q \times \Gamma \rightarrow \mathcal{P}(Q \times \Gamma \times \{L, R\})$ *is the transition relation,*
5. $q_0 \in Q$ *is the start state,*
6. $q_{accept} \in Q$ *is the accept state, and*
7. $q_{reject} \in Q$ *is the reject state, where* $q_{accept} \neq q_{reject}$.

Note that Δ is not a function anymore but a relation, reflecting the fact that an NTM does not have a single, uniquely defined next action but a choice between several next actions. In other words, for each state and symbol combination, there may be more than one appropriate next step, or none at all.

NTMs are powerful because of the asymmetrical input–output relation found in the way these machines compute. In order to have an NTM, M_N accepts one string w and it is enough to find just one branch b in the computation tree that accepts w.

It can be shown that an NTM can be simulated by a DTM, i.e. every NTM has an equivalent DTM [37]. However, simulating an NTM by a DTM may be at the cost of an exponential loss of efficiency [35]. Whether this loss is inherent to this "translation" between models or is just a consequence of our limited understanding of nondeterminism is the famous $\mathbf{P} \overset{?}{=} \mathbf{NP}$ problem [35].

Let us now formalize the idea of exponential loss of efficiency. To do so, we shall briefly review the main concepts behind complexity theory and NP-complete problems. These results will be used to show the importance of classical random walks in computer science as well as to provide the grounds for justifying the development of quantum walks and their use in algorithm development.

3.4   A QUICK TOUR ON ALGORITHMIC COMPLEXITY AND NP-COMPLETENESS

The performance of models of computation in the execution of an algorithm is a fundamental topic in the theory of computation. Since the quantification of resources (in our case, we focus on time) needed to find a solution to a problem is usually a complex process, we just estimate it. To do so, we use a form of estimation called **Asymptotic Analysis** in which we are interested in the maximum number of steps S_m that an algorithm must be run on large inputs. We do so by considering only the highest order term of the expression that quantifies S_m. For example, the function $F(n) = 18n^6 + 8n^5 - 3n^4 + 4n^2 - \pi$ has five terms, and the highest order term is $18n^6$. Since we disregard constant factors, we then say that f is asymptotically at most n^6. The following definition formalizes this idea.

Definition 3.4.1. Big O Notation. *Let $f, g : \mathbb{N} \to \mathbb{R}^+$. We say that $f(n) = O(g(n))$ if $\exists$ $\alpha, n_o \in \mathbb{N}$ such that $\forall n \geq n_o$*

$$f(n) \leq \alpha g(n).$$

So, $g(n)$ is an asymptotic upper bound for $f(n)$ (f is of the order of g). Bounds of the form n^β, $\beta > 0$ are called **polynomial bounds,** and bounds of the form 2^{n^γ}, $\gamma \in \mathbb{R}^+$ are called **exponential bounds.** $f(n) = O(g(n))$ means informally that f grows as g or slower.

Big O notation says that one function is asymptotically no more than another. To state that one function is asymptotically no less than another we use the Ω notation.

Definition 3.4.2. Ω Notation. *Let $f, g : \mathbb{N} \to \mathbb{R}^+$. We say that $f(n) = \Omega(g(n))$ if $\exists \alpha, n_o \in \mathbb{N}$ such that $\forall n \geq n_o$*

$$g(n) \leq \alpha f(n).$$

Finally, to say that two functions grow at the same rate we use the Θ notation.

Definition 3.4.3. Θ Notation. *Let $f, g : \mathbb{N} \to \mathbb{R}^+$. We say that $f(n) = \Theta(g(n))$ if $f(n) = O(g(n))$ and $f(n) = \Omega(g(n))$. Thus, $f(n) = \Theta(g(n))$ means that f and g have the same rate of growth.*

3.4.1 Algorithmic Complexity for DTMs

A DTM can be used to find a solution to a problem, so how efficiently can such a solution be found? As stated previously, we shall be interested in finding the fastest algorithms. Let us now introduce a few concepts needed to quantify the efficiency of an algorithm.

The time complexity of an algorithm A expresses its time requirements by giving, for each input length, the largest amount of time needed by A to solve a problem instance of that size.

Definition 3.4.4. Time Complexity Function for a DTM. *Let M be a DTM. We define $f : \mathbb{N} \to \mathbb{N}$ as the time complexity function of M, where $f(n)$ is the maximum number of steps that M uses on any input of length n.*

Definition 3.4.5. Time Complexity Class for DTMs. *Let $t : \mathbb{N} \to \mathbb{R}^+$ be a function. We define the time complexity class* **TIME(t(n))** *as the collection of all languages that are decidable by an $O(t(n))$ time DTM.*

Definition 3.4.6. Class P. *The class of languages that are decidable in polynomial time on a deterministic single-tape Turing machine is denoted by* **P** *and is defined as*

$$\mathbf{P} = \bigcup_k \mathbf{TIME}(n^k).$$

A *polynomial time* or *tractable algorithm* is an algorithm whose time complexity function is $O(p(n))$ for some polynomial function p, where n is used to denote the input length. Any algorithm whose time complexity function cannot be so bounded is called an *exponential time* or *intractable algorithm*. Tractable algorithms are considered as acceptable, in the sense that a satisfactory solution for a problem has been found. In contrast, intractable algorithms are usually solutions obtained by exhaustion, the so called brute-force method, and are not considered satisfactory solutions. As an example of a problem with no tractable algorithm associated up to date, we now define the satisfiability (SAT) problem.

Definition 3.4.7. The Satisfiability (SAT) Problem
Let $S = \{x_1, x_2, \ldots, x_n\}$ be a set of Boolean variables. A truth assignment for S is a function $t : S \rightarrow \{T, F\}$, for which if $t(x_i) = T$ we say that x_i is TRUE under t, and FALSE if $t(x_i) = F$. If x_i is a variable under S then x_i and $\bar{x}_i$ are literals over S. A clause over S is the disjunction of a set of literals over S (such as $x_1 \vee x_2 \vee \bar{x}_4$) and is satisfied by a truth assignment iff at least one of its members x_i is true under that assignment.

A collection C of clauses over S is satisfiable iff there exists some truth assignment for S that simultaneously satisfies all the clauses in C, i.e. C is a conjunction of disjunctions $C = \bigwedge_i [(\bigvee_j x_j)]$. Such a truth assignment is called a satisfying truth assignment for C.
INSTANCE: A set S of variables and a collection C of clauses over S.
QUESTION: Is there a satisfying truth assignment for C?

An example of the SAT problem follows. Suppose the existence of six variables $\{x_1, x_2, x_3, x_4, x_5, x_6\}$, their negations $\{\bar{x}_1, \bar{x}_2, \bar{x}_3, \bar{x}_4, \bar{x}_5, \bar{x}_6\}$, and the logical proposition

$$
\begin{aligned}
P = \ & (\bar{x}_1 \vee \bar{x}_4 \vee \bar{x}_5) \wedge (\bar{x}_2 \vee \bar{x}_3 \vee \bar{x}_4) \wedge (x_1 \vee x_2 \vee \bar{x}_5) \wedge (x_3 \vee x_4 \vee x_5) \wedge \\
& (x_4 \vee x_5 \vee \bar{x}_6) \wedge (\bar{x}_1 \vee \bar{x}_3 \vee \bar{x}_5) \wedge (x_1 \vee \bar{x}_2 \vee \bar{x}_5) \wedge (x_2 \vee \bar{x}_3 \vee \bar{x}_6) \wedge \\
& (\bar{x}_1 \vee \bar{x}_2 \vee \bar{x}_6) \wedge (x_3 \vee \bar{x}_5 \vee \bar{x}_6) \wedge (\bar{x}_1 \vee \bar{x}_2 \vee \bar{x}_4) \wedge (x_2 \vee x_3 \vee \bar{x}_4) \wedge \\
& (x_2 \vee x_5 \vee \bar{x}_6) \wedge (x_2 \vee \bar{x}_3 \vee \bar{x}_5) \wedge (\bar{x}_2 \vee \bar{x}_3 \vee \bar{x}_4) \wedge (x_2 \vee x_3 \vee x_6) \wedge \\
& (\bar{x}_1 \vee \bar{x}_2 \vee \bar{x}_3) \wedge (\bar{x}_1 \vee \bar{x}_4 \vee \bar{x}_5) \wedge (\bar{x}_3 \vee \bar{x}_4 \vee \bar{x}_6) \wedge (\bar{x}_4 \vee \bar{x}_5 \vee x_6) \wedge \\
& (\bar{x}_2 \vee x_3 \vee \bar{x}_6) \wedge (x_2 \vee x_5 \vee x_6) \wedge (x_3 \vee x_5 \vee \bar{x}_6) \wedge (\bar{x}_1 \vee x_3 \vee \bar{x}_6) \wedge \\
& (x_3 \vee \bar{x}_5 \vee x_6) \wedge (x_4 \vee x_5 \vee x_6) \wedge (x_1 \vee x_2 \vee \bar{x}_3).
\end{aligned}
$$

As can be seen, finding the solutions (if any) of even a modest SAT instance can become difficult quite easily. In fact, instance P has only one solution: $x_1 = 1$, $x_2 = 1$, $x_3 = 0$, $x_4 = 1$, $x_5 = 0$, $x_6 = 0$.

3.4.2   Algorithmic Complexity for NTMs

We start by offering the definitions of time complexity function and time complexity class for NTMs.

Definition 3.4.8.   Time Complexity Function for an NTM. *Let M_N be an NTM. We define $g : \mathbb{N} \to \mathbb{N}$ as the time complexity function of M_N, where $g(n)$ is the maximum number of steps that M_N uses on **any** branch of its computation on any input length n.*

Definition 3.4.9.   Time Complexity Class for NTMs. *Let $t : \mathbb{N} \to \mathbb{R}^+$ be a function. We define the time complexity class* **NTIME(t(n))**, *as the collection of all languages that are decidable by an $O(t(n))$ time nondeterministic Turing machine.*

For an NTM to accept string w it is enough to find just one branch b in its computation tree that accepts w. However, a practical problem with this definition is to find b as an NTM can have an infinite (or exponentially big) number of different branches. Therefore, a more operational method for doing nondeterministic computation is needed. An important discovery in the theory of computation is the fact that the complexities of many problems are linked by means of a concept called verifiability.

Definition 3.4.10.   Verifier. *A verifier for a language A is an algorithm V, where*

$$A = \{w | V \text{ accepts } (w, c) \text{ for some string } c\}.$$

We measure the time of a verifier only in terms of the length of w, so a polynomial-time verifier runs in polynomial time in the length of w. A language A is polynomially verifiable if it has a polynomial-time verifier. The string c, a certificate, is additional information needed by the verifier. For example, in the case of the SAT problem 3.4.7, c is the actual clause collection to be tested.

Note that a fundamental difference between an NTM (Def. 3.3.2) and a verifier is that an NTM *finds* solutions, while a verifier only checks whether a proposal is a solution or not.

We now proceed to define the most important class of languages in Computer Science.

Definition 3.4.11.   Class NP. *The class of languages that have polynomial-time verifiers is known as* **NP**.

What is the relation between the abstract model of an NTM and the concepts of verifiers and NP languages class? The answer is given in Theorem 1 and its proof can be found in [37].

Theorem 1. *A language is in* **NP** *if and only if it is decided by a nondeterministic polynomial-time Turing machine.*

3.4.3 P $\overset{?}{=}$ NP and NP-complete Problems

The problem **P** $\overset{?}{=}$ **NP** is a fundamental topic in the theory of computation. It is known that **P** $\subset$ **NP** as any polynomial language can be checked with a polynomial verifier. Also, it can be proved [62] that

Theorem 2. *If a problem $\zeta \in$ **NP** then $\exists$ a polynomial p such that ζ can be solved by a deterministic algorithm having time complexity $O(2^{p(n)})$.*

Due to Theorem 2 there is a widespread belief that **P** $\neq$ **NP** although no proof has been delivered and therefore **P** $\overset{?}{=}$ **NP** remains an open problem. A positive or negative but definite answer to **P** $\overset{?}{=}$ **NP** would provide computer scientists (and, in fact, all computer users) with an avalanche of answers with respect to the plausibility of finding exact solutions to problems that have been thought of as intractable for many years. The question **P** $\overset{?}{=}$ **NP** is so relevant in modern science that the Clay Research Institute, in remembrance of the famous 23 problem list delivered by Hilbert at the International Congress of Mathematicians at Paris in 1900, has included it in its Millennium Problems list[1].

There is a particular set of problems in **NP** that plays a key role in the theory of computation: **NP-complete** problems. In order to characterize this important set of problems we shall introduce the notion of polynomial transformations.

Definition 3.4.12. Polynomial Transformation. *A polynomial transformation from a language $L_1 \subset \Sigma_1^*$ to a language $L_2 \subset \Sigma_2^*$, denoted by $L_1 \propto L_2$, is a function f such that*
(1) there is a polynomial-time DTM that computes f and
(2) $\forall\, x \in \Sigma_1^*,\, x \in L_i \Leftrightarrow f(x) \in L_2$.

Definition 3.4.13. NP-Complete Languages and Problems. *A language L is* **NP-complete** *if $L \in$ **NP** and, for all other languages $L_i \in$ **NP** we find that $L_i \propto L$.*
Due to our capacity to go from problem instances to languages by means of encoding schemes, we can also say that a decision problem ζ is **NP-complete** *if $\zeta \in$ **NP** and, for all other decision problems $\zeta_i \in$ **NP** we find that $\zeta_i \propto \zeta$.*

There is a plethora of **NP-complete** problems. The first NP-complete problem (chronologically speaking) was found by Stephen Cook ([67], see also [62]), the SAT problem (Def. 3.4.7).

---

[1]http://www.claymath.org/millennium/.

Theorem 3. NP-Completeness of SAT problem. *SAT problem is NP-complete.*

Therefore, studying the properties of SAT is an important and active area of research, not only because a polynomial-time solution to SAT would imply $\mathbf{P} = \mathbf{NP}$, but also because SAT is used to model problems and procedures in several areas of applied computer science and engineering like Artificial Intelligence (e.g. [68]) and hardware verification (e.g. [69, 70]), using the following approach [43]:

1. Represent the problem in propositional logic.
2. Identify the proposition to be decided by satisfiability.
3. Solve the SAT problem.
4. Interpret the result in the original domain.

Surveys of algorithms for solving several variations and instances of SAT can be found in [43, 68, 71]. Also, good introductions to the vast field of computational complexity can be found in [37, 72–75].

3.5   PHYSICS AND THE THEORY OF COMPUTATION

Considerations about the physical properties of systems used to do computation and/or transmission of information have been studied for several decades. Consequently, physics and computer science have cross-fertilized each other for long time. As early as in the 1940s, in the beginning of the digital computer era, scientists wondered about the existence and quantification of the minimum amount of energy required to perform a computation. von Neumann, in a set of lectures delivered in 1949 [76], showed that "a minimum amount of energy required per elementary decision of a two-way alternative and the elementary transmittal of one unit of information" was close to kT, where k is Boltzmann's constant and T is the temperature of the system. Later on, Landauer studied the relationship between energy consumption and reversible computation (a computational step is reversible iff given the output of that step, its input is uniquely determined[2]). Among those results published by Landauer in [77], we have the following principle.

Landauer's principle. Suppose a computer erases a single bit of information. The amount of energy dissipated into the environment is at least $kT\ln 2$, where k is Boltzmann's constant and T is the temperature of the environment of the computer.

---

[2]For example, the logical operation **OR** is *not* reversible, while the operation **NOT** is indeed reversible.

Landauer's principle became a big motivation to do research in reversible computation. Among those works about reversible models of computation, we find [78–80].

Since evolution in quantum mechanics is reversible due to the use of unitary operators, the next step in the cross-fertilization between computer science and physics was to link quantum mechanics and computer science. Benioff introduced the notion of Quantum Turing Machines and proposed a quantum-mechanical model for the simulation of a classical computer ([26, Ch. 6] and [81–84]). Additionally, Feynman, in his traditional and celebrated style, lectured at MIT [5] about the fundamental capabilities and limitations of classical computers to simulate quantum systems. A gentle and concise introduction to this blend of physics, computer science and information theory, as well as Feynman's main ideas behind physics and computation can be found in [26].

In 1985, Deutsch made two key contributions in [66]: a design of a *Universal Quantum Turing Machine*, and a physics-oriented version of the Church–Turing thesis which he called "Church–Turing principle":

The Church–Turing principle [66]. Every finitely realizable physical system can be perfectly simulated by a universal model computing machine operating by finite means.

In Deutsch's words, the rationale behind the Church–Turing principle was "to reinterpret Turing's 'functions which would be naturally regarded as computable' as the functions which may, in principle, be computed by a real physical system. For it would surely be hard to regard a function 'naturally' as computable if it could not be computed in Nature, and conversely." The Universal Quantum Turing machine proposed in [66] was further developed and improved by Yao [85] and Bernstein and Vazirani [86].

We now define a Probabilistic Turing Machine and a Quantum Turing Machine.

Definition 3.5.1. *[28]* **Probabilistic Turing Machine**. *A Probabilistic Turing Machine (PTM) is a Nondeterministic Turing Machine which randomly chooses between the available transitions at each point according to a probability distribution. Thus, a PTM $M_N = (Q, \Sigma, \Gamma, \Delta, q_0, q_{accept}, q_{reject})$ is a 7-tuple where Q, Σ, Γ are all finite sets, $\mathcal{P}(Q \times \Gamma \times \{L, R\})$ is the power set of $Q \times \Gamma \times \{L, R\}$, and*

1. *Q is the set of states,*
2. *Σ is the input alphabet not containing the blank symbol $\sqcup$,*
3. *Γ is the tape alphabet, where $\sqcup \in \Gamma$ and $\Sigma \subset \Gamma$,*
4. *$q_0 \in Q$ is the start state,*
5. *$q_{accept} \in Q$ is the accept state, and*
6. *$q_{reject} \in Q$ is the reject state, where $q_{accept} \neq q_{reject}$,*

7. *The transition relation is given by* $\Delta : Q \times \Gamma \to \mathcal{P}(Q \times \Gamma \times \{L, R\} \times [0, 1])$, *so that for a given configuration* C_0, *each of its successor configurations* $C_1, C_2, \ldots, C_n$ *is assigned a probability* $p_1, p_2, \ldots, p_n$, *where* n *is the cardinality of* $\mathcal{P}(Q \times \Gamma \times \{L, R\} \times [0, 1])$ *and* $\sum_{i=1}^{n} p_i = 1$.

Definition 3.5.2. *[28]* **Quantum Turing Machine**. *A Quantum Turing Machine is defined analogously to a PTM but with a different transition relation. The transition relation includes the use of complex numbers which are the corresponding amplitudes of quantum states used for computation. A QTM is a 7-tuple* $M_N = (Q, \Sigma, \Gamma, \Delta, q_0, q_{accept}, q_{reject})$, *where* Q, Σ, Γ *are all finite sets,* $\mathcal{P}(Q \times \Gamma \times \{L, R\})$ *is the power set of* $Q \times \Gamma \times \{L, R\}$, *and*

1. Q *is the set of states,*

2. Σ *is the input alphabet not containing the blank symbol* $\sqcup$,

3. Γ *is the tape alphabet, where* $\sqcup \in \Gamma$ *and* $\Sigma \subset \Gamma$,

4. $q_0 \in Q$ *is the start state,*

5. $q_{accept} \in Q$ *is the accept state, and*

6. $q_{reject} \in Q$ *is the reject state, where* $q_{accept} \neq q_{reject}$,

7. *The transition relation is given by* $\Delta : Q \times \Gamma \to \mathcal{P}(Q \times \Gamma \times \{L, R\} \times \mathbb{C}_{[0,1]})$, *where* $\mathbb{C}_{[0,1]} = \{z \in \mathbb{C} | |z|^2 \leq 1\}$. *So, for a given configuration* C_0, *each of its successor configurations* $C_1, C_2, \ldots, C_n$ *is assigned an amplitude* $z_1, z_2, \ldots, z_n$, *where* n *is the cardinality of* $\mathcal{P}(Q \times \Gamma \times \{L, R\} \times \mathbb{C}_{[0,1]})$ *and* $\sum_{i=1}^{n} |z_i|^2 = 1$.

Quantum computation can be regarded as the study and development of methods that, by using quantum-mechanical properties, solve problems in finite time (from a different computational point of view, quantum computation is a sub-field of *unconventional models of computation* [87]). Quantum information can be defined as the field devoted to understanding how information is represented and communicated using quantum states. Due to the advances made over the last few years, both disciplines are now huge areas of research where diverse interests of several scientific communities can be found. Quantum walks is one of those interests, mainly contained in the sub-field of quantum algorithms.

CHAPTER 4

Classical Random Walks

A stochastic process is a system which evolves in time while undergoing chance fluctuations. We can describe such a system with a family of random variables $\{X_t\}$ where X_t measures, at time t, the property of the system which is of interest. If $t \in \mathbb{N}$ ($t \in \mathbb{R}^+ \cup \{0\}$) then $\{X_t\}$ is a discrete (continuous) stochastic process.

Among those stochastic processes relevant not only to mathematics itself but also to physics and computer science we find Discrete Random Walks (i.e. random walks on discrete spaces performed on discrete time steps) and Continuous Random Walks (i.e. random walks on discrete or continuous spaces performed on continuous time).

In this chapter we survey the statistical properties and computational applications of discrete random walks, followed by a summary of the basics of continuous random walks. Our focus is primarily on discrete random walks due to the extended use of this kind of stochastic processes in *stochastic algorithms* (also known as randomized algorithms), a branch of theoretical and applied computer science. Nonetheless, a discussion on continuous random walks is also necessary because the most impressive computational advantage of quantum walks published to date is an algorithm based on a continuous quantum walk [19]. To differentiate between random walks and their quantum counterparts, we shall refer to the former as (discrete or continuous) classical random walks and to the latter as (discrete or continuous) quantum walks.

4.1 PROBABILITY THEORY AND STOCHASTIC PROCESSES

In this section, based on [38, 39, 88–91], we provide some background results from probability theory and stochastic processes.

4.1.1 Discrete Random Variables and Distributions

Definition 4.1.1. Discrete Random Variable. *An experiment is a situation with a set of possible outcomes. Let us suppose we have an experiment with outcome space $\mathcal{E}$. A* **random variable** *(rv) is a real mapping $X : \mathcal{E} \to \mathbb{R}$ that is defined for all possible outcomes in S. A* **discrete random variable** *(drv) takes only a finite or countable infinite number of distinct values, i.e. $X : \mathcal{E} \to A \subset \mathbb{R}$ is a drv iff $\#(A) \leq \aleph_0$. The expression $X = x_i$ is shorthand for $X(e_i) = x_i, \forall\, e_i \in \mathcal{E},\, x_i \in A$.*

Definition 4.1.2.   Probability distribution for a drv. *Let $X : \mathcal{E} \to A$ be a drv. Since the outcomes of an experiment are uncertain in general, we associate with each outcome $x_i \in A$ a probability $p(x_i)$, where $p(x_i) = \Pr(X = x_i)$. The numbers $p(x_i)$ are called a* **probability distribution of** *X iff (i) $p(x_i) \geq 0$ and (ii) $\sum_{x_i \in A} p(x_i) = 1$.*

Definition 4.1.3.   Expectation value and variance. *The expectation value μ of a drv X, also known as* **mean***, is defined as $E[X] = \sum_i x_i p(x_i)$. More generally, the expectation value of any function $g(X)$ of X is given by $E[f(X)] = \sum_i f(x_i) p(x_i)$. The* **variance** *$V[X]$ of a distribution, also written as σ^2, is given by $V[X] = E[(X - \mu)^2] = \sum_i (x_i - \mu)^2 p(x_i)$. The square root of the variance is known as the* **standard deviation** *and is denoted by σ.*

If X, Y are two drv and $a, b \in \mathbb{R}$, the following propositions can be proved [38]:

$$E[aX + bY] = a E[X] + b E[Y], \tag{4.1}$$
$$V[X] = E[X^2] - (E[X])^2. \tag{4.2}$$

If X, Y are independent drvs, then

$$V[aX + bY] = a^2 V[X] + b^2 V[Y]. \tag{4.3}$$

Definition 4.1.4.   Bernoulli distribution. *The Bernoulli distribution, denoted $\mathcal{B}(\theta)$, is used as a model for experiments which have only two outcomes: success with probability θ, and failure with probability $1 - \theta$. If X is $\mathcal{B}(\theta)$ then $X = 1$ if success and $X = 0$ if failure. It is a well-known fact that if X is $\mathcal{B}(\theta)$ then $\mu_X = \theta$ and $\sigma^2 = \theta(1 - \theta)$.*

Definition 4.1.5.   Binomial distribution. *The binomial distribution, denoted $\text{Bin}(n, p)$, describes experiments that consist of a number of independent identical trials with two possible outcomes: success with probability p and failure with probability $q = 1 - p$. So, the random variable $X =$ "number of successes" can take any value from $\{1, 2, \ldots, n\}$ and its distribution is described by the binomial distribution. If X is $\text{Bin}(n, p)$ then the probability $p(r)$ of obtaining r successes from n trials is given by $p(r) = \binom{n}{r} p^r q^{n-r}$.*

Definition 4.1.6.   Geometric distribution. *The geometric distribution describes experiments that consist of a number of independent trials, each having a probability p, which are performed until a success occurs. If we let X be the number of trials required then the probability of getting a successful result after n trials is given by $P(X = n) = (1 - p)^{n-1} p$. If X is geometrically distributed then $E[X] = \frac{1}{p}$.*

Theorem 1. Markov's inequality. *Let X be a drv that takes only non-negative values, then*

$$P(X \geq a) \leq \frac{E[X]}{a}.$$

The mean and the variance of a drv X, although important quantities, do not contain all the information about the distribution of that variable[1]. One way to completely characterize the probability distribution of a drv X is to use the *moments* of X.

4.1.2   Moments and Generating Functions

Definition 4.1.7.   Moments of a drv. *We define the kth moment of a drv $X : \mathcal{E} \rightarrow A$ as $\mu_k = E(X^k) = \sum_{j=1}^{\infty}(x_j)^k p(x_j)$ where $x_i \in A$, and under the assumption that the sum converges. It can be proved [39] that, in terms of its moments, the mean and the variance of a drv X are given by*

$$\mu = \mu_1 \tag{4.4}$$
$$\sigma^2 = \mu_2 - \mu_1^2. \tag{4.5}$$

Generating functions are a powerful and compact mathematical concept (power series) to encode information about sequences. In order to produce a *moment generating function* for a drv X, let us define the function

$$e^{tX} = \sum_{k=0}^{\infty} \frac{t^k}{k!} X^k. \tag{4.6}$$

By Def. 4.1.3, we have

$$E(e^{tX}) = \sum_{j=0}^{\infty} e^{tx_j} p(x_j). \tag{4.7}$$

Thus

$$E(e^{tX}) = E\left(\sum_{k=0}^{\infty} \frac{t^k}{k!} X^k\right) = \sum_{k=0}^{\infty} \frac{t^k}{k!} E(X^k) = \sum_{k=0}^{\infty} \frac{t^k}{k!} \mu_k.$$

Combining Eqs. (4.6) and (4.7), we obtain

$$g(t) = E(e^{tX}) = \sum_{j=0}^{\infty} e^{tx_j} p(x_j) = \sum_{k=0}^{\infty} \frac{t^k}{k!} \mu_k. \tag{4.8}$$

---

[1]It is possible to find two different probability distributions p_1 and p_2 corresponding to two different drvs X_1 and X_2 with the same mean and variance, i.e. $\mu_{X_1} = \mu_{X_2}$ and $\sigma_{X_1}^2 = \sigma_{X_2}^2$.

Definition 4.1.8.   Moment generating function. *The function*

$$g(t) = E(e^{tX}) = \sum_{j=0}^{\infty} e^{tx_j} p(x_j) = \sum_{k=0}^{\infty} \frac{t^k}{k!} \mu_k$$

is known as the **moment generating function for** X.

Theorem 2.   *Let $g(t)$ be a moment generating function for drv $X \Rightarrow \frac{d^n}{dt^n} g(t) \Big|_{t=0} = \mu_n$.*

Proof.   By induction we find that $\frac{d^n}{dt^n} g(t) = \sum_{k=n}^{\infty} \frac{k(k-1)(k-2)\cdots(k-(n+1))}{k!} \mu_k t^{k-n}$.

Since $k(k-1)(k-2)\cdots(k-(n+1)) = \frac{k!}{(k-n)!} \Rightarrow \frac{d^n}{dt^n} g(t) = \sum_{k=n}^{\infty} \frac{k!}{(k-n)!k!} \mu_k t^{k-n}$.

We define $\alpha_i = \frac{(n+i)!}{i!(n+i)!} \Rightarrow \frac{d^n}{dt^n} g(t) \Big|_{t=0} = \left[\mu_n + \sum_{i=n+1}^{\infty} \alpha_i \mu_i t^i \right]_{t=0} = \mu_n.$ □

The binomial distribution is widely used in the study of classical random walks. In the following theorem, we compute the moment generating function of a drv $X \, \text{Bin}(n, p)$.

Theorem 3.   *Let X be Bin(n,p) $\Rightarrow g_X(t) = (e^t p + q)^n$, with $q = 1 - p$.*

Proof.   By definition $g(t) = \sum_{j=0}^{\infty} e^{tx_j} p(x_j) \Rightarrow g_X(t) = \sum_{j=0}^{n} \binom{n}{j} e^{tj} p^j q^{n-j}$.

By the binomial theorem $\sum_{j=0}^{n} \binom{n}{j} e^{tj} p^j q^{n-j} = (e^t p + q)^n$. □

Note that, if X is Bin(n, p) then

$$\mu_1 = \frac{d}{dt} g(t) \Big|_{t=0} = ne^t p(e^t p + q)^{n-1} \Big|_{t=0} = np \quad \text{and} \quad \mu_2 = \frac{d^2}{dt^2} g(t) \Big|_{t=0} = n(n-1)p^2 + np.$$

Therefore, if a drv X is Bin(n, p) then its mean and variance are given by

$$\mu = \mu_1 = np, \tag{4.9a}$$

and

$$\sigma^2 = \mu_2 - \mu_1^2 = np(1 - p). \tag{4.9b}$$

The following theorem establishes the convergence and uniqueness properties of moment generating functions, and its proof can be found in [39].

Theorem 4. *Let X be a drv with finite range $\{x_1, x_2, \ldots x_n\}$, distribution function p and moments $\mu_k \Rightarrow$*

(i) The moment series $g(t) = \sum_{k=0}^{\infty} \frac{\mu_k t^k}{k!}$ converges for all t to an infinitely differentiable function $g(t)$.

(ii) The moment series $g(t) = \sum_{k=0}^{\infty} \frac{\mu_k t^k}{k!}$ is uniquely determined by p and conversely.

Finally, with respect to generating functions, let us focus on the particular but important case in which a drv X takes values $x_j \in \mathbb{N} \cup \{0\}$. In this case, it is useful to have an alternative definition of a moment generating function.

Definition 4.1.9. Ordinary generating functions. *Let $X : \mathcal{E} \to \mathbb{N} \cup \{0\}$ be a drv and $g(t)$ be its moment generation function. Since $g(t) = \sum_{j=0}^{\infty} e^{t x_j} p(x_j) = \sum_{j=0}^{\infty} e^{t j} p(j)$ then $g(t)$ is a polynomial in e^t. Let us perform the variable change $z = e^t$ and define the function h as*

$$h(z) = \sum_{j=0}^{\infty} p(j) z^j.$$

*The function $h(z)$ is called the **ordinary generating function** for X. Note that $h(1) = g(0) = 1$, $h'(1) = g'(0) = \mu_1$, and $h''(1) = g''(0) - g'(0) = \mu_2 - \mu_1$. An ordinary generating function is also simply called a **probability generating function (pgf)**.*

We use Def. 4.1.9 in the following result on Bernoulli distributions.

Theorem 5. *Let X be Bernoulli distributed with parameter $\theta \Rightarrow h_X(z) = 1 - \theta + \theta z$.*

Proof. $p(X = 0) = 1 - \theta$ and $p(X = 1) = \theta \Rightarrow h(z) = \sum_{j=0}^{\infty} z^j p(j) = (1-\theta) z^0 + \theta z^1 = 1 - \theta + \theta z$. $\square$

Now we introduce another powerful mathematical tool to study classical random walks in both lines and graphs: Markov chains.

4.1.3 Markov Chains

Definition 4.1.10. Markov chain. *Let $\{X_\alpha | \alpha \in \mathbb{N} \cup \{0\}\}$ be a set of discrete random variables and S be a system defined by the state space $\{s_\beta | \beta \in \mathbb{N} \cup \{0\}\}$. X_n is defined as the state of a system S at time n, so we say that S is in state s_i at time n iff $X_n = s_i$.*

*The sequence $\{X_\alpha\}$ is said to form a **Markov chain** with initial distribution λ and transition matrix $\mathbf{P}$ if each time S is in state s_i there is some fixed probability p_{ij} that it will be in state s_j, and*

p_{ij} *does not depend upon which states the chain was in before the current state. In other words,*

$$P(X_{n+1} = s_j | X_n = s_{j-1})$$
$$= P(X_{n+1} = s_j | X_n = s_{j-1} \wedge X_{n-1} = s_{j-2} \wedge \ldots X_1 = s_1 \wedge X_0 = s_0) = p_{ij}$$

where the initial state s_0 is drawn from the initial distribution λ. The values p_{ij} are called the **transition probabilities** *of the Markov chain and they satisfy $p_{ij} \geq 0$ and $\sum_i p_{ij} = 1$, as transition probabilities must conform a probability distribution. Transition matrices are also called right stochastic matrices, i.e. matrices for which the sum of the elements of each row is equal to 1.*

The following lemma and theorem (corresponding proofs can be found in [39]) show the relationship between the time evolution of a Markov chain and its transition matrix **P**.

Lemma 1. *Let* **P** *be the transition matrix of a Markov chain. The ij-th entry $\mathbf{p}_{ij}^n$ of the matrix $\mathbf{P}^n$ gives the probability that the Markov chain, starting in state s_i, will be in state s_j after n steps.*

Theorem 6. *Let* **P** *be the transition matrix of a Markov chain, and let $\vec{\lambda}$ be the probability row vector which represents the initial distribution $\lambda \Rightarrow$ the probability that the chain is in state s_i after n steps is the i-th entry of the* **row** *vector $\vec{\lambda}^{(n)}$, given by*

$$\vec{\lambda}^{(n)} = \vec{\lambda} \mathbf{P}^n.$$

We now present an example of a stochastic system and its transition matrix.

Example 4.1.1.   Drunkard's walk. A man is frequent visitor of a pub which is located five blocks from his home. If he is in any corner between home and the pub he walks to the left or to the right (i.e. toward home or the pub) with equal probability. Also, if he reaches either home or the pub he stays there.

The behavior of this man can be modeled by a Markov chain with state space $S = \{1, 2, 3, 4, 5, 6\}$ being state $s_1 = 1$ home and state $s_6 = 6$ the pub, i.e. s_0 and s_5 are the absorbing states of this walk. The transition matrix is then

$$\mathbf{P} = \begin{pmatrix} 1 & 0 & 0 & 0 & 0 & 0 \\ 1/2 & 0 & 1/2 & 0 & 0 & 0 \\ 0 & 1/2 & 0 & 1/2 & 0 & 0 \\ 0 & 0 & 1/2 & 0 & 1/2 & 0 \\ 0 & 0 & 0 & 1/2 & 0 & 1/2 \\ 0 & 0 & 0 & 0 & 0 & 1 \end{pmatrix}.$$

We are interested in studying several cases of Markov chains. The most important case is that in which it is possible to visit every state of the system S with no other constraint apart from the probabilities defined by the powers of the transition matrix $\mathbf{P}$. The following definitions and theorems provide the grounds to characterize such Markov chains.

Definition 4.1.11.   Stationary probability distribution. *Let* $\mathbf{P}$ *be a transition matrix. A stationary probability distribution is a row vector* $\overrightarrow{\pi}$ *that satisfies* $\lim_{n\to\infty} \overrightarrow{\lambda_0} \mathbf{P}^n = \overrightarrow{\pi}$.

Theorem 7.   *Let* $\mathbf{P}$ *be a transition matrix and* π *a stationary probability distribution* $\Rightarrow \overrightarrow{\pi} = \overrightarrow{\pi} \mathbf{P}$.

Definition 4.1.12.   Irreducibility of a Markov chain. *Let* $\{X_\alpha\}$ *be a Markov chain with state space* $S = \{s_\beta\}$ *and transition matrix* $\mathbf{P}$. $\{X_\alpha\}$ **is irreducible** *if* $\forall s_i, s_j \in S \; \exists \, t \in \mathbb{N}$ *such that* $\mathbf{p}_{ij}^t > 0$.

So, a Markov chain is irreducible if it is possible to visit any state. In order to avoid any "visiting pattern," we shall impose another condition on Markov chains, that of *aperiodicity*.

Definition 4.1.13.   Aperiodicity of a Markov Chain. *Let* $\{X_\alpha\}$ *be a Markov chain with state space* $S = \{s_\beta\}$ *and transition matrix* $\mathbf{P}$. $\{X_\alpha\}$ *is aperiodic if* $\forall s_i, s_j \in S \Rightarrow gdc\{t \in \mathbb{N} | \mathbf{p}_{ij}^t > 0\} = 1$.

Definition 4.1.14.   Ergodic Markov chains. *A Markov chain is* **ergodic** *if it is irreducible and aperiodic.*

We are now ready to mention the most important theorem of Markov chains.

Theorem 8.   Fundamental theorem of Markov chains *[39]. An ergodic Markov chain has a unique stationary distribution* $\overrightarrow{\pi}$ *. For any initial probability distribution* $\overrightarrow{\lambda}$ *we have* $\overrightarrow{\lambda} \mathbf{P}^t \to \overrightarrow{\pi}$ *as* $t \to \infty$.

So, if we let an ergodic Markov chain run for long enough, it will eventually lose all memory of where it started and will reach some fixed distribution $\overrightarrow{\pi}$ over its state space $S = \{s_\beta\}$. Therefore, the unique stationary distribution $\overrightarrow{\pi}$ of an ergodic Markov chain is **independent** of the initial probability distribution $\overrightarrow{\lambda}$. This fact will be an important factor to differentiate between classical random walks and quantum walks.

4.2   CLASSICAL DISCRETE RANDOM WALKS: RESULTS AND APPLICATIONS

The previous section will now be used to develop some important results of classical discrete random walks on a line, on a circle, and on a graph. Those results will be employed in this lecture to present some applications of classical random walks in computer science, as well as to show the differences between this kind of stochastic processes and discrete quantum walks.

Classical discrete random walks were first thought as stochastic processes with no relation to algorithm development, thus besides to classical references on random walks like [92–94], it is necessary to scan articles and a few books in order to find relevant material. Therefore, in addition to the references mentioned at the beginning of this chapter, we have used [40–42] for this section.

4.2.1   Classical Discrete Random Walks on a Line

A classical discrete random walk on a line is a particular kind of stochastic process. The simplest classical random walk on a line consists of a particle (the walker) jumping to either left or right depending on the outcomes of a probability system (the coin) with (at least) two mutually exclusive results, i.e. the particle moves according to a probability distribution.

The generalization to discrete random walks on spaces of higher dimensions (graphs) is straightforward. An example of a discrete random walk on a graph is a particle moving on a lattice where each node has six vertices, and the particle moves according to the outcomes produced by tossing a dice. In fact, a classical random walk on a line is also a random walk on a graph $G = (V, E)$ with $|V| = 2$. Classical random walks on graphs can be seen as Markov chains [23, 89]. Furthermore, if the random walk is aperiodic and irreducible then it has a stationary distribution (Theorem 8).

Unrestricted Classical Discrete Random Walk on a Line

Let $\{Z_n\}$ be a stochastic process which consists of the path of a particle which moves along an axis with steps of one unit at time intervals also of one unit (Fig. 4.1). At any step, the particle has a probability p of going to the right and $q = 1 - p$ of going to the left. Compute the probability of finding the particle in position k after n steps. $\{Z_n\}$ has time parameter space $\mathbb{N}$, discrete state space $\mathbb{Z}$, and the starting point is $Z_0 = 0$. Each step is an independent drv X with distribution $\mathrm{pr}(X = 1) = p$ and $\mathrm{pr}(X = -1) = q$. After n steps we can see that $Z_n = \sum_{i=1}^{n} X_i$.

We are interested in finding the value $P_k^n = \mathrm{pr}(Z_n = k | Z_0 = 0)$, so we define a new drv Y_i

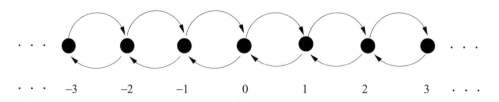

FIGURE 4.1: An unrestricted classical discrete random walk on a line. The probability of going to the right is p and the probability of going to the left is $q = 1 - p$.

$$Y_i = \begin{cases} 1 & \text{if } x_i = 1 \\ 0 & \text{if } x_i = -1. \end{cases}$$

Each drv $Y_i = \frac{1}{2}(X_i + 1)$ is an independent Bernoulli trial (Def. 4.1.4) with a probability of success p. We use $\{Y_i\}$ to define a drv T_n that represents the "number of successes"

$$T_n = \sum_{k=1}^{n} Y_i = \frac{1}{2}(Z_n + n).$$

T_n is Bin(n, p) (Def. 4.1.5) $\Rightarrow \text{pr}(Z_n = k | Z_0 = 0) = \text{pr}(T_n = \frac{1}{2}(Z_n + n) = \frac{1}{2}(k + n)) \Rightarrow$

$$\text{pr}(Z_n = k | Z_0 = 0) = \begin{cases} \binom{n}{\frac{1}{2}(k+n)} p^{\frac{1}{2}(k+n)} q^{\frac{1}{2}(n-k)}, & \frac{1}{2}(k + n) \in \mathbb{N} \cup \{0\} \\ 0, & \text{otherwise.} \end{cases} \tag{4.10}$$

Since T_n is Bin(n, p) then $E[T_n] = np$ and $V[T_n] = npq$. So, using Eq. (4.1) we find

$$E[Z_n] = E[2T_n - n] = n(p - q). \tag{4.11}$$

Similarly, using Eq. (4.3)

$$V[Z_n] = V[2T_n - n] = 4npq. \text{ In other words, } V[Z_n] = O(n). \tag{4.12}$$

Classical Discrete Random Walk on a Line With Two Absorbing Barriers

We analyze the case of the path of a particle which moves along a *finite* axis with steps of one unit at time intervals of one unit. The axis has **absorbing boundaries** $-a$ and b, i.e. if the particle reaches either $-a$ or b it remains there. As in the previous case, the particle has a probability p of going to the right and $q = 1 - p$ of going to the left, and each step is independent of every other step.

Let $\{Z_n\}$ be the stochastic process that models the path of this particle, with time parameter space $\mathbb{N}$ and state space $\{-a, -a + 1, \ldots, -1, 0, 1, 2, \ldots, b - 1, b\}$. We are interested in computing the probability of $Z_n = -a$ *before* $Z_n = b$ (see Fig. 4.2).

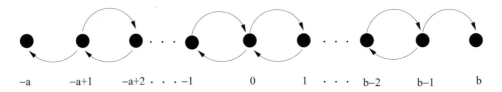

FIGURE 4.2: Classical discrete random walk on a line with two absorbing barriers. The probability of going to the right is p and the probability of going to the left is $q = 1 - p$, except for the extreme sites in which the walker is absorbed with probability 1.

This problem is known as the Gambler's ruin problem because one can think of it as two gamblers A and B with corresponding capitals of $\$a$ and $\$b$. A and B play a game in which each play results in A winning $\$1.00$ from B with probability p or B winning $\$1.00$ from A with probability q. We want to know the probability that gambler A is in ruin.

Let us define the drv

$$Y_i = \begin{cases} 1 & \text{if A is eventually ruined, i.e. } Z_n = -a \text{ before } Z_n = b \\ 0 & \text{otherwise.} \end{cases}$$

Y is $\mathcal{B}(\theta)$ (Def. 4.1.4), so the pgf of Y given that the walk starts in state i (Theorem 5) is given by $h(z)_{Y_{(i)}} = 1 - (1 - z)\theta_i$ and we want to find θ_0, i.e. we want $\text{pr}(Y = 1 | Z_0 = 0)$. Using techniques for solving difference equations, we find that

$$\theta_0 = \begin{cases} \frac{b}{a+b} & \lambda = \frac{p}{q} = 1 \\ \frac{\lambda^b - 1}{\lambda^b - \lambda^{-a}} & \lambda = \frac{p}{q} \neq 1. \end{cases} \tag{4.13}$$

Similarly, the probability that A is triumphant is given by

$$\phi_0 = \begin{cases} \frac{a}{a+b} & \lambda = \frac{p}{q} = 1 \\ \frac{1 - \lambda^a}{1 - \lambda^{a+b}} & \lambda = \frac{p}{q} \neq 1. \end{cases} \tag{4.14}$$

We prove that the game will eventually end simply by showing that A will either lose or win with probability 1: $\theta_0 + \phi_0 = 1$.

Classical Discrete Random Walk on a Line With One Absorbing Barrier

This problem can be thought as a variation of the Gambler's ruin problem, with gambler B having unlimited capital (B could be, for example, a casino). Therefore, we define a stochastic process $\{Z_n\}$ that models the path of a particle moving along an axis. Z_n has time parameter space $\mathbb{N}$ and state space $\{-a, -a+1, \ldots, -1, 0\} \bigcup \mathbb{N}$. As before, the particle has a probability p of going to the right and $q = 1 - p$ of going to the left, and each step is independent of every other step (see Fig. 4.3). We are interested in computing the probability $\text{pr}(Z_n = -a | Z_0 = 0)$. This probability can be found by computing the limit

$$\text{pr}(Z_n = -a | Z_0 = 0) = \lim_{b \to \infty} \theta_0 = \begin{cases} 1 & \text{if } p \leq q \\ \left(\frac{q}{p}\right)^a & \text{if } p > q. \end{cases} \tag{4.15}$$

So, if B has unlimited capital and unless A has a success probability higher than that of his/her opponent, it is certain that A will be eventually ruined.

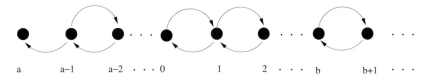

a   a−1   a−2 · · · 0   1   2 · · · b   b+1 · · ·

FIGURE 4.3: Classical discrete random walk on a line with one absorbing barrier. The walker can be absorbed in node a. The probability of going to the right is p and the probability of going to the left is $q = 1 - p$. In node a, the probability of being absorbed is equal to 1.

4.2.2   Classical Discrete Random Walks on a Graph

A graph is a symbolic representation of a network and of its connectivity. Of particular importance in computer science is the relationship between graphs, Markov chains, and classical discrete random walks.

Definition 4.2.1.   Graph. *A* **graph** *$G = (V, E)$ is a set V of vertices v_i connected by edges $(v_k, v_l) \in E$. We define $|V|$ as the total number of vertices and $|E|$ as the total number of edges of G. The* **degree** *of a vertex is the number of edges of that vertex.*

A graph is **connected** *if there is a path connecting every pair of vertices. A graph is* **bipartite** *if its set of vertices can be divided into two disjoint sets with two vertices of the same set never sharing an edge, and* **non-bipartite** *otherwise. If $\forall (u, v) \in E \ni (v, u) \in E \Rightarrow G$ is* **undirected***.*

A graph can be represented by its **adjacency matrix** *$A = (a_{ij})$, which is a matrix with rows and columns labeled by graph vertices, with entries $a_{ij} = 1$ or 0 according to whether vertices i and j are linked by an edge or not.*

Graphs that encode the structure of a group are called **Cayley graphs**. Group theory is a branch of mathematics widely used in several fields of science and engineering (quantum physics and control theory, for example). Thus, Cayley graphs are a vehicle for translating mathematical structures of scientific and engineering problems into forms amenable to algorithm development for scientific computing.

Definition 4.2.2.   Cayley graph. *Let G be a finite group, and let $S = \{s_1, s_2, \ldots, s_k\}$ be a generating set for G. The* **Cayley graph** *of G with respect to S has a vertex for every element of G, with an edge from g to $gs \ \forall g \in G$ and $s \in S$.*

Cayley graphs are k-regular, that is, each vertex has degree k. Cayley graphs have more structure than arbitrary Markov graphs and their properties are often used in algorithm development [95].

Graphs and Markov chains can be put in an elegant framework which turns out to be very useful for the development of algorithmic applications.

Let $G = (V, E)$ be a connected, undirected graph with $|V| = n$ and $|E| = m$. G induces a Markov chain M_G if the states of M_G are the vertices of G, and $\forall\, u, v \in V$

$$p_{uv} = \begin{cases} \frac{1}{d(u)} & \text{if } (u, v) \in E \\ 0 & \text{otherwise,} \end{cases}$$

where $d(u)$ is the degree of vertex u. Since G is connected, then M_G is irreducible and aperiodic [23], therefore M_G has a unique stationary distribution (Theorem 8).

Theorem 9. *Let G be a connected, undirected graph with n nodes and m edges, and let M_G be its corresponding Markov chain. Then, M_G has a unique distribution*

$$\vec{\pi} = (d(v_i)/2m)$$

for all components v_i of $\vec{\pi}$.

Note that Theorem 9 holds even when the distribution $\{d(v_i)\}$ is not uniform. In particular, the stationary distribution of an undirected and connected graph with n nodes, m edges, and constant degree $d(v_i) = r\ \forall\ v_i \in G$, i.e. a Cayley graph, is $\vec{\pi} = (r/2m)$, the uniform distribution.

We have established the relationship between Markov chains and graphs. We now proceed to define the concepts that make discrete random walks on graphs useful in computer science. We shall begin by formally describing a random walk on a graph: let G be a graph. A random walk, starting from a vertex $u \in V$ is the random process defined by

s = u

repeat

    choose a neighbor v of u according to a certain probability distribution P

    u = v

until (stop condition)

So, we start at a node v_0 and, if at t-th step we are at a node v_t, we move to a neighbor of v_t with probability given by probability distribution P. It is common practice to make $P_{uv} = \frac{1}{d(v_t)}$, where $d(v_t)$ is the degree of vertex v_t. Examples of discrete random walks on graphs are a classical random walk on a circle or on a three-dimensional mesh.

We now introduce several measures to quantify the performance of discrete random walks on graphs. These measures play an important role in the quantitative theory of random walks, as well as in the application of this kind of Markov chains in computer science.

Definition 4.2.3. Hitting time. *The hitting time H_{ij} is the expected number of steps before node j is visited, starting from node i.*

Definition 4.2.4.   Mixing rate. *The mixing rate is a measure of how fast the discrete random walk converges to its limiting distribution. The mixing rate can be defined in many ways, depending on the type of graph we want to work with. We use the definition given in [41].*

If the graph is non-bipartite then $p_{ij}^t \to d_j/2m$ as $t \to \infty$, and the mixing rate is given by

$$\mu = \lim_{t \to \infty} \sup \max \left| p_{ij}^{(t)} - \frac{d_j}{2m} \right|^{1/t}.$$

As it is the case with the mixing rate, the **mixing time** can be defined in several ways. Basically, the notion of mixing time comprises the number of steps one must perform a classical discrete random walk before its distribution is close to its limiting distribution.

Definition 4.2.5.   Mixing time *[96]. Let M_G be an ergodic Markov chain which induces a probability distribution $P_u(t)$ on the states at time t. Also, let $\vec{\pi}$ denote the limiting distribution of M_G. The mixing time τ_ϵ is then defined as*

$$\tau_\epsilon = \max_u \min_t \{t | t \geq T \Rightarrow || P_u(t) - \vec{\pi} || < \epsilon\}$$

where $|| P_u(t) - \vec{\pi} ||$ is a standard distance measure. For example, we could use the total variation distance, defined as $|| P_u(t) - \vec{\pi} || = \frac{1}{2} \sum_i |P_{u_i}(t) - \pi_i|$. Thus, the mixing time is defined as the first time t such that $P_u(t)$ is within distance ϵ of $\vec{\pi}$ at all subsequent time steps $t \geq T$, irrespective of the initial state.

Calculating mixing times is a difficult task. Consequently, there are several strategies to compute mixing times. Among them we find the **coupling time strategy**, which consists of considering two discrete random walks on a Markov chain. By starting one of the random walks from the stationary distribution and bounding the time for the two chains to collide, we can compute bounds on the mixing time of the random walk. What does it mean to make two chains collide? That means that both chains will end up hitting the same nodes with the same probability. To formalize this concept, let us present the following theorem.

Theorem 10.   *Let P and Q be two probability distributions with $P_x^{(t)}$ and $Q_x^{(t)}$ the probabilities of hitting node x at time $t \Rightarrow |P - Q| \leq 2pr(P_x^{(t)} \neq Q_x^{(t)})$.*

So, the computation of the mixing time of a Markov chain by means of the coupling strategy consists of the following steps: 1. Compute the limiting distribution of the Markov chain. 2. Compute the time it takes to obtain the following equality: $P_x^{(t)} = \pi_x$, where π_x is the probability of hitting node x according to the Markov chain's limiting distribution $\vec{\pi}$. This step is usually equivalent to computing the hitting time of the Markov chain for a certain node. The key question is: how many steps n does it take to hit node k?

We now present the mixing times of several classical discrete random walks.

Mixing Time of an Unrestricted Classical Discrete Random Walk on a Line

It has been shown in Eq. (4.10) that, for an unrestricted classical discrete random walk on a line with $p = q = \frac{1}{2}$, the probability of finding the walker in position k after n steps is given by

$$\text{pr}(Z_n = k \mid Z_0 = 0) = \begin{cases} \binom{n}{\frac{1}{2}(k+n)} \frac{1}{2^n}, & \frac{1}{2}(k+n) \in \mathbb{N} \cup \{0\} \\ 0, & \text{otherwise.} \end{cases}$$

Using Stirling's approximation $n! \approx \sqrt{2\pi n}\left(\frac{n}{e}\right)^n$ and after some algebra, we find

$$\text{pr}(Z_n = k \mid Z_0 = 0) = \frac{1}{2^n}\binom{n}{\frac{1}{2}(k+n)} \approx \sqrt{\frac{2n}{\pi^2(n^2 - k^2)}} \frac{n^n}{(n+k)^{\frac{n+k}{2}}(n-k)^{\frac{n-k}{2}}}. \tag{4.16}$$

We know that Eq. (4.10) is a binomial distribution, thus it makes sense to study the mixing time in two different vertex populations: $k \ll n$ and $k \approx n$ (the first population is mainly contained under the bell-shape part of the distribution, while the second can be found along the tails of the distribution). In both cases, we shall find the expected hitting time by calculating the inverse of Eq. (4.16) (this is the expected time of the geometric distribution given in Def. 4.1.6).

Case $k \ll n$. Since

$$\sqrt{\frac{2n}{\pi^2(n^2 - k^2)}} \frac{n^n}{(n+k)^{\frac{n+k}{2}}(n-k)^{\frac{n-k}{2}}} \approx \sqrt{\frac{2n}{\pi^2 n^2}} \frac{n^n}{n^{n/2}n^{n/2}} = \frac{c}{\sqrt{n}} \Rightarrow$$

$$\text{Hitting time } H_{0,k} = O(\sqrt{n}). \tag{4.17}$$

Case $k \approx n$. Let $n - k = C_1$ and $n^2 - k^2 = C_2$, where C_1 and C_2 are small integer numbers. Since

$$\sqrt{\frac{2n}{\pi^2(n^2 - k^2)}} \frac{n^n}{(n+k)^{\frac{n+k}{2}}(n-k)^{\frac{n-k}{2}}} \approx \sqrt{\frac{2n}{\pi C_2}} \frac{n^n}{2^n n^n C_1^{C_1/2}} = \frac{1}{2^n}\sqrt{\frac{2n}{\pi C_1^{C_1} C_2}} \Rightarrow$$

$$\text{Hitting time } H_{0,k} = O(2^n). \tag{4.18}$$

Thus, the hitting time for a given vertex k of an n-step unrestricted classical discrete random walk on a line depends on which region vertex k is located in. If $k \ll n$ then it will take

$\sqrt{n}$ steps to reach k, on average. However, if $k \approx n$ then it will take an exponential number of steps to reach k, as one would expect from the properties of the binomial distribution. So, if we use these hitting times to get a qualitative picture of the corresponding mixing time, i.e. the time it takes to a binomial distribution of n steps to "look like" (that is, to be ϵ-close to) a binomial distribution computed after an infinite (or, being rigorous, a very large) number of steps, we find that the computation of such mixing time is not straightforward. It seems that more analysis and new methods for computing mixing times are needed in order to study unrestricted classical random walks, particularly within the framework of algorithm development (in fact, to the best of our knowledge, there is only a very limited number of publications, among them [40], that work on the properties of classical random walks on infinite graphs).

Mixing Time of a Classical Discrete Random Walk on a Line With Two Reflecting Barriers

Let $\{Z_n\}$ be a stochastic process which consists of the path of a particle which moves along a finite axis with steps of one unit at time intervals also of one unit. The axis has n different position sites. At any step, the particle has a probability p of going to the right and $q = 1 - p$ of going to the left, except for the case in which the particle is sitting on an extreme point $Z_t = 0$ or $Z_t = n - 1$. If the particle is on $Z_t = 0$ ($Z_t = n - 1$) at time t then the particle will move to $Z_{t+1} = 1$ ($Z_{t+1} = n - 2$) at time $t + 1$ with probability 1 (see Fig. 4.4). According to Theorem 9, $\{Z_n\}$ has a stationary distribution given by

$$\vec{\pi} = \frac{1}{n+1},$$ (4.19)

and a hitting time $H_{0,n}$ given by [41]

$$H_{0,n} = O(n^2).$$ (4.20)

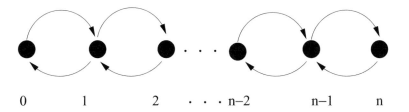

$$0 \qquad 1 \qquad 2 \quad \cdot \cdot \cdot \; n{-}2 \qquad n{-}1 \qquad n$$

FIGURE 4.4: Classical discrete random walk on a line with two reflecting barriers. The probability of going to the right is p and the probability of going to the left is $q = 1 - p$. In the extremes, the probability of "bouncing" is equal to 1.

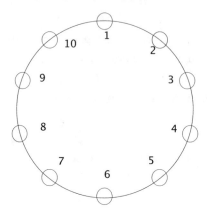

FIGURE 4.5: Classical discrete random walk on a ten-nodes circle.

The stationary distribution from Eq. (4.19) is independent of p and q because this result is a particular case of Theorem 9, which in turn is independent of specific values of p and q.

Mixing Time of a Classical Discrete Random Walk on a Circle

The definitions of discrete random walks on a circle and on a line with two reflecting barriers are very similar. In fact, the only difference is the behavior of the extreme nodes.

Let $\{Z_n\}$ be a stochastic process which consists of the path of a particle which moves along a circle with steps of one unit at time intervals also of one unit. The circle has n different position sites (for an example with ten nodes, see Fig. 4.5). At any step, the particle has a probability p of going to the right and $q = 1 - p$ of going to the left. If the particle is on $Z_t = 0$ at time t then the particle will move to $Z_{t+1} = 1$ with probability p and to $Z_{t+1} = n - 1$ with probability q. Similarly, if the particle is on $Z_t = n - 1$ at time t then at time $t + 1$ the particle will go to $Z_{t+1} = 0$ with probability p and to $Z_{t+1} = n - 2$ with probability q.

According to Theorem 9, the Markov chain defined by $\{Z_n\}$ has a stationary distribution given by

$$\vec{\pi} = \frac{1}{n}, \tag{4.21}$$

and a hitting time $H_{0,n}$ given by [41]

$$H_{0,n} = O(n^2). \tag{4.22}$$

4.3   STOCHASTIC ALGORITHMS BASED ON CLASSICAL DISCRETE RANDOM WALKS

Our motivation for studying random walks is to understand how such stochastic processes can be used to build algorithms. In this section we show how discrete random walks are used to solve two different instances of the SAT problem (Def. 3.4.7).

Algorithms that use stochastic processes to find a solution, i.e. procedures that make random choices during execution, are known as **randomized algorithms**. In our chapter on the theory of computation, we defined a Probabilistic Turing Machine (Def. 3.5.1) as an NTM which randomly chooses between the available transitions at each point according to a given probability distribution. Thus, a randomized algorithm is a PTM.

In this section we present two randomized algorithms based on classical discrete random walks: the first one solves 2-SAT in polynomial time, and the second is the most efficient algorithm (though still exponential) known for solving 3-SAT. Some more uses of hitting times for developing on-line algorithms can be found in [97].

The **SAT** problem is a key element in the theory of computation (Theorem 3). Let us remark that in the definition of SAT problem (Def. 3.4.7) there is no constraint neither in the number of clauses nor in the number of literals per clause. Thus, in order to make SAT amenable to algorithmic analysis we define a variation with a limited number of literals per clause:

Definition 4.3.1.   The k-SAT Problem. *Let $S = \{x_1, x_2, \ldots, x_n\}$ be a set of Boolean variables and C be a collection of clauses over S where each clause has k literals, i.e. C is a conjunction of disjunctions $C = \bigwedge_i [(\bigvee_{j=1}^{k} x_j)]$.*
INSTANCE: A set S of variables and a collection of clauses over S.
QUESTION: Is there a satisfying truth assignment for C?

4.3.1   2-SAT

In 2-SAT we have a proposition of the form $C = \bigwedge_i [(\bigvee_{j=1}^{2} x_j)]$ and we are interested in finding a set of values for the variables $x_1, x_2, \ldots, x_n$ such that $C = TRUE$. In [35, 75], Papadimitriou proposed the following randomized algorithm for the solution of 2-SAT.

Algorithm 1. Randomized algorithm for 2-SAT.
Input: a proposition $C = \bigwedge_i [(\bigvee_{j=1}^{2} x_j)]$ with a total number of n variables $x_1, x_2, \ldots, x_n$.
Objective: To determine whether C is satisfiable or not.

Steps of the algorithm
1. Start with any truth assignment T

2. **repeat** r times
3.   **if** there is no unsatisfied clause then
4.   Reply "formula is satisfiable"
5.   Halt
6.   **else**
7.   Take any unsatisfied clause
8.   Pick any of these two literals and flip it, updating T
9. After r repetitions reply "Formula is probably unsatisfiable"

Algorithm 1 is randomized because in step 8 we make a random choice of the literal whose value will be changed. We obtain additional randomness by randomly selecting an initial truth assignment T (step 1) and an unsatisfied clause (step 7). We focus on step 8.

In order to analyze algorithm 1, let clause C be satisfiable by truth assignment T_s. We define E_i as the **expected number of repetitions** of step 8 until a satisfying truth assignment is found, *under the assumption that our starting truth assignment T differs from T_s in exactly i values*, i.e. $d(T, T_s) = i$. Iterating on step 8 can be seen as a classical discrete random walk on a line in which positions on the line correspond to the actual distance between T and T_s, and the walker moves to the left or right with probabilities α and β, respectively (if variables are picked up uniformly at random then $\alpha = \beta = \frac{1}{2}$). The following theorem states the performance of this randomized algorithm for 2-SAT.

Theorem 11.   Papadimitrious's solution to 2-SAT. *Suppose that a discrete random walk algorithm (algorithm 1) with $r = 2n^2$ is applied to any satisfiable instance of 2-SAT with n variables. Then the probability that a satisfying truth assignment will be discovered is at least $\frac{1}{2}$.*

Proof. We begin executing algorithm 1 with an initial truth assignment T and distance $d(T, T_s) = i$. In terms of the discrete random walk picture, we begin our walk on position i and we want to compute E_i.

If the walker moves to the left (i.e. we get closer to truth assignment T_s by flipping the right variable) then this new scenario can be described as a discrete random walk starting at position $i - 1$ and having expected number of steps $E_{i-1} = E_i - 1$ or, equivalently, $E_i = E_{i-1} + 1$. If α is the probability of going to the left and β to the right, we then obtain the following expression:

$$E_i = 1 + \alpha E_{i-1} + \beta E_{i+1}. \tag{4.23}$$

Eq. (4.23) is a difference equation constrained by the following conditions:
1. $E_0 = 0$. The number of expected steps when we have a satisfying assignment is zero.
2. $E_n = E_{n-1} + 1$. If we arrive at position n we must make one step to the left (i.e. we have

reached a limit and therefore we must return) and that scenario is equivalent to starting a discrete random walk in position $n - 1$ with expected number of steps $E_{n-1} = E_n - 1$.

A random walk constrained by the previous conditions is known as *a discrete random walk with one absorbing barrier ($E_0 = 0$) and one reflecting barrier ($E_n = E_{n-1} + 1$)*. Strictly speaking, Eq. (4.23) should be the inequality $E_i \leq 1 + \alpha E_{i-1} + \beta E_{i+1}$, the reason being that C could have more than one satisfying truth assignment that could also be found during the computation of algorithm 1. So, Eq. (4.23) represents the worst case as is standard practice in algorithm performance analysis. Using standard methods for difference equations [43] with $\alpha = \beta = \frac{1}{2}$, we find that Eq. (4.23) has solution $E_i = 2in - i^2$, therefore

$$E_n = n^2.$$

Thus, our expected number of steps is n^2 regardless our starting point. Finally, using Markov's inequality (Theorem 1) we find that

$$P(X \geq 2n^2) \leq \frac{n^2}{2n^2} = \frac{1}{2}.$$

$\square$

A detailed analysis of Theorem 11 is given in [43], along with a proof that algorithm 1 together with the techniques used in Theorem 11 are feasible for 2-SAT only, as solutions for 3-SAT and beyond are exponentially complex.

4.3.2 3-SAT

Despite its polynomial time performance, the discrete random walk solution from the previous subsection is not the most efficient method known (a linear time solution was proposed in [98]). However, the scenario changes when dealing with more complicated problems like 3-SAT, in which case the algorithm proposed by Schöning in [24] provides the technique used to achieve the best performance up to date for solving k-SAT (the best-known upper bound for 3-SAT is given in [99] and is an improved version of Schöning's proposal).

The algorithm proposed in [24] is given in the following lines:

Algorithm 2. Randomized algorithm for k-SAT.
Input: a proposition $C = \bigwedge_i [(\bigvee_{j=1}^k x_j)]$ with a total number of n variables $x_1, x_2, \ldots, x_n$.
Objective: To determine whether C is satisfiable or not.
1. Start with any truth assignment T
2. **repeat** $3n$ times
3.   **if** there is no unsatisfied clause then
4.   Reply "formula is satisfiable"

5.   Halt
6.   **else**
7.   Take any unsatisfied clause
8.   Pick one of the k literals in the clause and flip it, updating T

Let us suppose that C is satisfiable by T_s. The purpose of [24] is to estimate the probability p of reaching T_s with initial truth assignment T, by executing algorithm 2 under the constraints that will be explained in the following lines. Once probability p is known it is also possible to estimate the expected number of times $E_t = \frac{1}{p}$ that algorithm 2 should be executed in order to reach T_s (a sequence of independent repetitions of algorithm 2 with "success probability" p can be described by a geometrically distributed drv X (Def. 4.1.6)) if $E_t = \frac{1}{p}$ then the complexity of the algorithm is within a polynomial factor of $\frac{1}{p}$.

In contrast to Papadimitriou's solution [75], in this case the discrete random walk is performed only a limited number of times ($3n$) and algorithm 2 is repeated approximately E_t times (this is called the "restart effect," which has a positive impact in the performance of algorithm 2 [43]). Additionally and under the assumption that our starting truth assignment T differs from T_s in exactly j values, i.e. $d(T, T_s) = j$, the random walk in algorithm 2 is allowed to make only $i \le j$ "wrong" steps, i.e. steps of the form $d(T, T_s) = k \to d(T, T_s) = k + 1$. So, the random walk is expected to take $j + 2i$ steps in order to reach state 0. Note that $j + 2i \le n + 2n = 3n$ is a necessary condition (otherwise algorithm 2 would never reach state 0).

By calculating the number of paths which take the walker from j to 0 with i steps in the "wrong" direction and following the mathematical details provided in [24], it is possible to conclude that the probability p is given by $p \ge (\frac{1}{2}(1 + \frac{1}{k-1}))^n$. Therefore, the complexity of algorithm 2 is within a polynomial factor of $(2(1 - \frac{1}{k}))^n$.

4.4   CLASSICAL CONTINUOUS RANDOM WALKS

So far in this chapter we have focused on studying several properties of random walks on discrete spaces and discrete time steps. We now extend our study to random walks defined in continuous spaces.

Continuous random walks are a subfield of Markov processes, a branch of mathematics extensively used in many fields of physics and engineering. Furthermore, the concepts of continuous Markov processes have been extended into theoretical and applied computer science like Machine Learning (for example [100]) and some links between continuous Markov processes, quantum mechanics and computer science have been developed over the last few years (for example [19, 20]).

In this section we shall briefly study the definitions of two different kinds of continuous random walks: those that are performed on continuous graphs and continuous time, and those defined on discrete graphs and continuous time.

Let $G = (V, E)$ be a regular isotropic graph, i.e. a graph with constant vertex degree and with equal transition probability between neighboring sites. Now, suppose that the distance d_{ij} between nodes of the graph decreases, i.e. $d_{ij} \to 0$. Also, suppose that the transition time between nodes is now a real non-negative value rather than a discrete variable, i.e. $t \in \mathbb{R}^+ \cup \{0\}$ instead of $t \in \mathbb{N} \cup \{0\}$. Finally, let us assume that the transition probability is equal for any pair of nodes, i.e. the process is isotropic.

Let us focus on the computation of transition probability density P, that is, the probability that the walker goes to node x assuming that the walker was located at $x_0 = 0$ at time $t = 0$. It can be proved that P obeys the diffusion equation

$$\frac{\partial P}{\partial t} = D\nabla^2 P, \tag{4.24}$$

where ∇ is the Laplacian operator and D is the diffusion coefficient.

Now let us suppose that $t \in \mathbb{R}^+ \cup \{0\}$ but the nodes of the graph are discrete. In this case ∇^2 is replaced by the Laplacian of the graph

$$L_{ij} = \begin{cases} d_i, & i = j \\ -1, & (i, j) \in G \\ 0, & \text{otherwise.} \end{cases} \tag{4.25}$$

As we shall see in Chapter 6, Eqs. (4.24) and (4.25) are closely related to the mathematical formulation of continuous quantum walks.

CHAPTER 5

Quantum Walks

Quantum walks are quantum counterparts of classical random walks. As shown in our previous chapter, classical random walks have been successfully adopted to develop classical algorithms. Since one of the main topics in quantum computation is the creation of quantum algorithms which are faster than their classical counterparts, there has been a huge interest in understanding the properties of quantum walks over the last few years. In addition to their usage in computer science, the study of quantum walks is relevant to building methods in order to test the "quantumness" of emerging technologies for the creation of quantum computers.

Quantum walks is a new research topic. Although some authors have selected the name "quantum random walk" to refer to quantum phenomena [101–103] and, in fact, in the seminal work by Feynman's about quantum-mechanical computers [104] we find a proposal that could be interpreted as a (continuous) quantum walk [105], it is generally accepted that the first paper with quantum walks as its main topic was published in 1993 by Aharonov et al. [106]. Thus, the links between classical random walks and quantum walks, as well as the utility of quantum walks in computer science, are two fresh and open areas of research. As we have seen in the previous chapter, there is a theory of classical random walks on finite graphs that, although still far from complete, it has been fruitful in algorithm development. In order to fully capitalize quantum walks in computer science, we still need to do more work on performance measures to compare quantum and classical performance, as well as to produce new ideas on how to benefit from applying quantum walks in algorithm design (as we shall see in the following chapter, some algorithms based on quantum walks have already been proposed, but only one algorithm based on a continuous quantum walk has rendered an exponential speedup with respect to its classical counterparts).

Two models of quantum walks have been suggested:
- The first model, called **discrete quantum walks**, consists of two quantum-mechanical systems (a walker and a coin) as well as an evolution operator which is applied to both systems only in discrete time steps although, as we shall see in the following lines, there are some proposals about avoiding the use of coins [107, 108] in discrete quantum walks. Discrete quantum walks with coins are also called **coined discrete quantum walks**.

- In the second model, named **continuous quantum walks**, the evolution operator of the system can be applied at any time.

In both cases, the quantum walk is performed on discrete graphs (a summary of the basics of both kinds of quantum walks can be found in [44]).

The original idea behind the construction of quantum algorithms was to start by initializing a set of qubits and then to apply (one of more) evolution operators several times *without performing intermediate measurements*, as measurements were meant to be performed only at the end of the computational process. By doing so, quantum interference would cause a behavior radically different from that of a classical algorithm.

Not surprisingly, quantum algorithms based on quantum walks have been designed using the same strategy: initialize qubits, apply evolution operators and measure only to calculate the final outcome of the algorithm. Indeed, this method has proved itself very useful for building several remarkable algorithms [44, 109]. However, it has recently been reported that performing (partial) measurements on a quantum walk may lead to interesting mathematical properties for algorithm development, like the "top hat" probability distribution [110, 111].

Quantum walks is a new tool expected to play a major role in the field of quantum algorithms, and a number of benefits of employing such walks in algorithm development are already known, as we shall see in the following chapter.

The rest of this chapter is organized as follows. We begin by delivering a detailed analysis of the unrestricted discrete quantum walk on a line with a Hadamard coin operator, followed by an examination of a discrete quantum walk on a line with a general coin, and the effect of using several kinds of coins in quantum walks. We then proceed to review some results on quantum walks on a line with boundaries, followed by a summary of properties and main results on quantum walks on graphs. We then briefly review some studies that focus on the transition between classical to quantum walks and vice versa, as well as on the "quantumness" of quantum walks and the role of entanglement in discrete quantum walks.

The second and last parts of this chapter provide a succinct introduction to continuous quantum walks, move ahead with an analysis of the randomness of a quantum walk and then focus on how continuous and discrete quantum walks are connected.

5.1   QUANTUM WALK ON A LINE

Discrete quantum walks on a line (DQWL) is the most studied model of discrete quantum walks. As its name suggests, this kind of quantum walks are performed on graphs $G = (V, E)$ of degree $|V| = 2$ (Def. 4.2.1). Studying DQWL is important in quantum computation for several reasons, including:

1. DQWL can be used to build quantum walks on more sophisticated structures like circles or general graphs.

2. DQWL is a simple model that can be exploited to explore, find and understand relevant properties of quantum walks for the development of quantum algorithms.

3. DQWL can be employed to test the quantumness of experimental realizations of quantum computers.

In [112], Meyer made two contributions to the study of DQWL while studying the models of Quantum Cellular Automata (QCA) and Quantum Lattice Gases:

1. He proposed a model of quantum dynamics that would be used later on to analytically characterize DQWL.

2. He showed that a quantum process in which, at each time step, a quantum particle (the walker) moves in superposition both to left and right with equal amplitudes, is physically impossible in general, the only exception being the trivial motion in a single direction.

In order to perform a discrete DQWL with non-trivial evolution, it was proposed in [96, 113] to use an additional quantum system: a coin. Thus, a DQWL comprises two quantum systems, **coin** and **walker**, along with a coin unitary operator (to toss a coin) and a conditional shift operator (to displace the walker either to the left or right depending on the accompanying coin state component).

In a different perspective, Patel et al. proposed in [107] to eliminate the use of coins by rearranging the Hamiltonian operator associated with the evolution operator of the quantum walk (however, there is a price to be paid on the translation invariance of the quantum walk). Moreover, Hines and Stamp have proposed the development of quantum walk Hamiltonians [114] in order to reflect the properties of potential experimental realizations of quantum walks in their mathematical structure.

Motivated by [107], Hamada et al. [115] wrote a general setting for QCA, developed a correspondence between DQWL and QCA, and applied this connection to show that the quantum walk proposed in [107] could be modeled as a QCA. The relationship between QCA and quantum walks has been indirectly explored by Meyer [112]. Additionally, Konno et al. [116] have studied the relationship between quantum walks and cellular automata, and it has been shown by van Dam [117] that it is possible to build a quantum cellular automaton capable of universal computation. Studying the relationship between QCA and quantum walks may lead to interesting computability properties of quantum walks.

The rest of this section is organized as follows. First, we review the mathematical structure of a coined DQWL. We then proceed to study in detail the properties of a discrete quantum walk on an infinite line, followed by the cases of one and two absorbing boundaries. We finish with a study on the impact of using multiple coins on quantum walks on a line.

5.1.1   Structure of a Coined DQWL

The main components of a coined DQWL are a walker, a coin, evolution operators for both walker and coin, and a set of observables:

Walker and Coin: The walker is a quantum system living in a Hilbert space of infinite but countable dimension $\mathcal{H}_p$. It is customary to use vectors from the canonical (computational) basis of $\mathcal{H}_p$ as "position sites" for the walker. So, we denote the walker as $|\text{position}\rangle \in \mathcal{H}_p$ and affirm that the canonical basis states $|i\rangle_p$ that span $\mathcal{H}_p$, as well as any superposition of the form $\sum_i \alpha_i |i\rangle_p$ subject to $\sum_i |\alpha_i|^2 = 1$, are valid states for $|\text{position}\rangle$. The walker is usually initialized at the "origin," i.e. $|\text{position}\rangle_{\text{initial}} = |0\rangle_p$.

The coin is a quantum system living in a two-dimensional Hilbert space $\mathcal{H}_c$. The coin may take the canonical basis states $|0\rangle$ and $|1\rangle$ as well as any superposition of these basis states. Therefore, $|\text{coin}\rangle \in \mathcal{H}_c$ and a general normalized state of the coin may be written as $|\text{coin}\rangle = a|0\rangle_c + b|1\rangle_c$, where $|a|^2 + |b|^2 = 1$.

The total state of the quantum walk resides in $\mathcal{H}_t = \mathcal{H}_p \otimes \mathcal{H}_c$. So far, only product states of $\mathcal{H}_t$ have been used as initial states, that is, $|\psi\rangle_{\text{initial}} = |\text{position}\rangle_{\text{initial}} \otimes |\text{coin}\rangle_{\text{initial}}$.

Evolution Operators: The evolution of a quantum walk is divided into two parts that closely resemble the behavior of a classical random walk. In the classical case, chance plays a key role in the evolution of the system. This is evident in the following example: we first toss a coin (either biased or unbiased) and then, depending on the coin outcome, the walker moves one step either to the right or to the left.

In the quantum case, the equivalent of the previous process is to apply an evolution operator to the coin state followed by a conditional shift operator to the total quantum system. The purpose of the coin operator is to render the coin state in a superposition, and the randomness is introduced by performing a measurement on the system after both evolution operators have been applied to the total quantum system several times.

Among coin operators, customarily denoted by $\hat{C}$, the Hadamard operator (Eq. (2.4)) has been extensively employed. For convenience we show it again in the following equation:

$$\hat{H} = \frac{1}{\sqrt{2}}(|0\rangle_c \langle 0| + |0\rangle_c \langle 1| + |1\rangle_c \langle 0| - |1\rangle_c \langle 1|). \tag{5.1}$$

For the conditional shift operator use is made of a unitary operator that allows the walker to go one step forward if the accompanying coin state is one of the two basis states (e.g. $|0\rangle$), or one step backwards if the accompanying coin state is the other basis state ($|1\rangle$). A suitable conditional shift operator has the form

$$\hat{S} = |0\rangle_c \langle 0| \otimes \sum_i |i+1\rangle_p \langle i| + |1\rangle_c \langle 1| \otimes \sum_i |i-1\rangle_p \langle i|. \tag{5.2}$$

Consequently, the operator on the total Hilbert space is $\hat{U} = \hat{S} \cdot (\hat{C} \otimes \hat{\mathbb{I}}_p)$ and a succinct mathematical representation of a quantum walk after t steps is

$$|\psi\rangle_t = (\hat{U})^t |\psi\rangle_{\text{initial}}, \tag{5.3}$$

where $|\psi\rangle_{\text{initial}} = |\text{position}\rangle_{\text{initial}} \otimes |\text{coin}\rangle_{\text{initial}}$.

Observables: Several advantages of quantum walks over classical random walks are a consequence of interference effects between coin and walker after several applications of $\hat{U}$ (other advantages come from quantum entanglement between walker(s) and coin(s) as well as partial measurement and/or interaction of coins and walkers with the environment). However, we must perform a measurement at some point in order to know the outcome of our walk. To do so, we define a set of observables according to the basis states that have been used to define coin and walker.

There are several ways to extract information from the composite quantum system. For example, we may first perform a measurement on the coin using the observable

$$\hat{M}_c = \alpha_0 |0\rangle_c \langle 0| + \alpha_1 |1\rangle_c \langle 1|. \tag{5.4}$$

A measurement must then be performed on the position states of the walker by using the operator

$$\hat{M}_p = \sum_i a_i |i\rangle_p \langle i|. \tag{5.5}$$

We show in Fig. 5.1 the probability distributions of two 100-steps DQWL. Coin and shift operators for both quantum walks are given by Eqs. (5.1) and (5.2), respectively. The DQWLs from plots (a) and (b) have corresponding initial quantum states $|0\rangle_c \otimes |0\rangle_p$ and $|1\rangle_c \otimes |0\rangle_p$. The first evident property of these quantum walks is the skewness of their probability distributions, as well as the dependence of the symmetry of such a skewness from the coin initial quantum state ($|0\rangle$ for plot (a) and $|1\rangle$ for plot (b)). This skewness comes from constructive and destructive interference due to the minus sign included in Eq. (5.1). Also, we note a quasi-uniform behavior in the central area of both probability distributions, approximately in the interval $[-70, 70]$. Finally, we note that regardless their skewness, both probability distributions cover the same number of positions (in this case, even positions from -100 to 100. If the quantum walk had

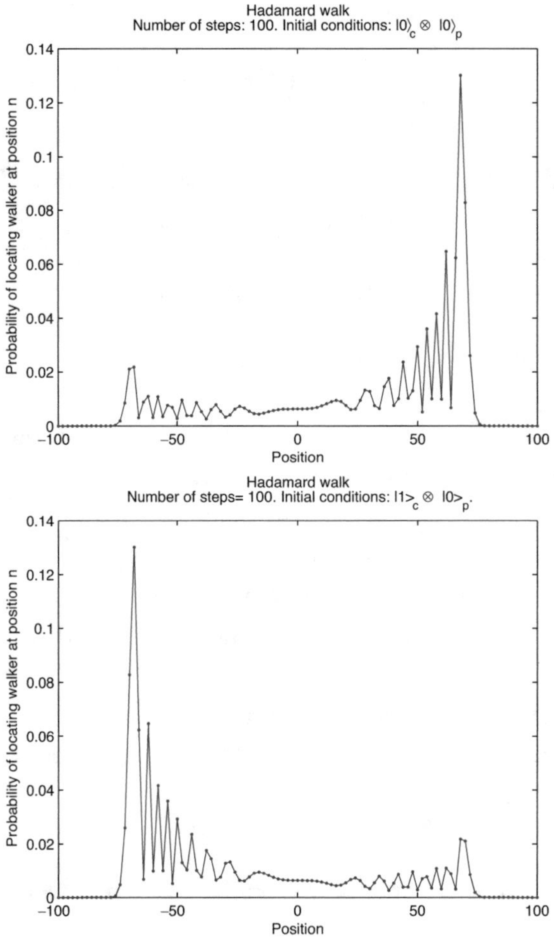

FIGURE 5.1: Probability distributions of 100 steps DQWLs using coin and shift operators given by Eqs. (5.1) and (5.2), respectively. Plot (a) corresponds to a DQWL with total initial quantum state $|0\rangle_c \otimes |0\rangle_p$, while plot (b) had total initial quantum state $|1\rangle_c \otimes |0\rangle_p$. Two interesting properties of these quantum walks is the skewness of corresponding probability distributions, along with the dependence of the symmetry of such skewness from the coin initial state.

been performed an odd number of times, then only odd position sites could have non-zero probability). A few steps of a DQWL are presented immediately after Eq. (5.7).

5.1.2   Analysis of Quantum Walks on an Infinite Line

Two approaches have been extensively used to study DQWL:

1. Schrödinger approach. In this case, we take an arbitrary component $|\psi\rangle_n = (\alpha|1\rangle_c +$

$\beta |0\rangle_c) \otimes |n\rangle_p$ of the quantum walk, the tensor product of coin and position components for a certain walker position. $|\psi\rangle_n$ is then Fourier-transformed in order to get a closed form of the coin amplitudes. Then, standard tools of complex analysis are managed to calculate the statistical properties of the probability distribution computed from corresponding coin amplitudes.

2. Combinatorial approach. In this method we compute the amplitude for a particular position component $|n\rangle_p$ by summing up the amplitudes of all the paths which begin in the given initial condition and end up in $|n\rangle_p$. This approach can be seen as using a discrete version of path integrals.

More recently, Fuss et al. have proposed an analytic description of probability densities and moments for the one-dimensional quantum walk on a line [118]. Moreover, Feldman and Hillery [119] have proposed an alternative formulation of discrete quantum walks based on scattering theory.

In the following lines we review both Schrödinger and combinatorial approaches to analyze the Hadamard walk, a specific but very powerful DQWL with coin and shift operators given by Eqs. (5.1) and (5.2), respectively. Later on we show how the Hadamard walk is related to the more general case of a DQWL with arbitrary coin operator.

Schrödinger Approach for the Hadamard Walk
The analysis of DQWL properties using the Discrete Time Fourier Transform (DTFT) and methods from complex analysis was first made by Nayak and Vishwanath [113], followed by Ambainis et al. [96], Košík [120], and Carteret et al. [121, 122]. Following [96, 113], a quantum walk on an infinite line after t steps can be written as $|\psi\rangle = (\hat{U})^t |\psi\rangle_{\text{initial}}$ (Eq. (5.3)) or, alternatively, as

$$\sum_k [a_k |0\rangle_c + b_k |1\rangle_c]|k\rangle_p \tag{5.6}$$

where $|0\rangle_c, |1\rangle_c$ are the coin state components and $|k\rangle_p$ are the walker state components. For example, let us suppose we have

$$|\psi\rangle_0 = |0\rangle_c \otimes |0\rangle_p \tag{5.7}$$

as the quantum walk initial state, with Eqs. (5.1) and (5.2) as coin and shift operators, respectively. Then, the first three steps of this quantum walk can be written as

$$|\psi\rangle_1 = \frac{1}{\sqrt{2}}|0\rangle_c |1\rangle_p + \frac{1}{\sqrt{2}}|1\rangle_c |-1\rangle_p ,$$

$$|\psi\rangle_2 = \left(\frac{1}{2}|0\rangle_c + 0|1\rangle_c\right)|2\rangle_p + \left(\frac{1}{2}|0\rangle_c + \frac{1}{2}|1\rangle_c\right)|0\rangle_p + \left(0|0\rangle_c - \frac{1}{2}|1\rangle_c\right)|-2\rangle_p \,,$$

and

$$|\psi\rangle_3 = \left(\frac{1}{2\sqrt{2}}|0\rangle_c + 0|1\rangle_c\right)|3\rangle_p + \left(\frac{1}{\sqrt{2}}|0\rangle_c + \frac{1}{2\sqrt{2}}|1\rangle_c\right)|1\rangle_p +$$
$$\left(\frac{-1}{2\sqrt{2}}|0\rangle_c + 0|1\rangle_c\right)|-1\rangle_p + \left(0|0\rangle_c + \frac{1}{2\sqrt{2}}|1\rangle_c\right)|-3\rangle_p \,.$$

We now define

$$\Psi(n, t) = \begin{pmatrix} \Psi_R(n, t) \\ \Psi_L(n, t) \end{pmatrix} \tag{5.8}$$

as the two component vector of amplitudes of the particle being at point n and time t or, in operator notation

$$|\Psi(n, t)\rangle = \Psi_L(n, t)|1\rangle + \Psi_R(n, t)|0\rangle. \tag{5.9}$$

We shall now analyze the behavior of a Hadamard walk at point n after $t + 1$ steps. We begin by applying the Hadamard operator given by Eq. (5.1) to those coin state components in positions $n - 1$, n, and $n + 1$:

$$\hat{H}(|\Psi(n - 1, t)\rangle + |\Psi(n, t)\rangle + |\Psi(n + 1, t)\rangle)$$
$$= \frac{1}{\sqrt{2}}(|\Psi_L(n - 1, t)\rangle|0\rangle + |\Psi_R(n - 1, t)\rangle|0\rangle - |\Psi_L(n + 1, t)\rangle|1\rangle + |\Psi_R(n + 1, t)\rangle|1\rangle$$
$$- |\Psi_L(n - 1, t)\rangle|1\rangle + |\Psi_R(n - 1, t)\rangle|1\rangle + |\Psi_L(n + 1, t)\rangle|0\rangle + |\Psi_R(n + 1, t)\rangle|0\rangle$$
$$+ |\Psi_L(n, t)\rangle|0\rangle + |\Psi_R(n, t)\rangle|0\rangle - |\Psi_L(n, t)\rangle|1\rangle + |\Psi_R(n, t)\rangle|1\rangle). \tag{5.10}$$

Now, we apply the shift operator given by Eq. (5.2) to Eq. (5.10):

$$\hat{U}(\hat{H}(|\Psi(n - 1, t)\rangle + |\Psi(n, t)\rangle + |\Psi(n + 1, t)\rangle))$$
$$= \frac{1}{\sqrt{2}}(|\Psi_L(n, t)\rangle|0\rangle + |\Psi_R(n, t)\rangle|0\rangle - |\Psi_L(n, t)\rangle|1\rangle + |\Psi_R(n, t)\rangle|1\rangle$$
$$- |\Psi_L(n - 2, t)\rangle|1\rangle + |\Psi_R(n - 2, t)\rangle|1\rangle + |\Psi_L(n + 2, t)\rangle|0\rangle + |\Psi_R(n + 2, t)\rangle|0\rangle$$
$$- |\Psi_L(n - 1, t)\rangle|1\rangle + |\Psi_R(n - 1, t)\rangle|1\rangle + |\Psi_L(n + 1, t)\rangle|0\rangle + |\Psi_R(n + 1, t)\rangle|0\rangle). \tag{5.11}$$

The bold font amplitude components of Eq. (5.11) are the amplitude components of $|\Psi(n, t + 1)\rangle$, which can be written in matrix notation as

$$\Psi(n, t + 1) = \begin{pmatrix} \frac{-1}{\sqrt{2}} & \frac{1}{\sqrt{2}} \\ 0 & 0 \end{pmatrix} \Psi(n + 1, t) + \begin{pmatrix} 0 & 0 \\ \frac{1}{\sqrt{2}} & \frac{1}{\sqrt{2}} \end{pmatrix} \Psi(n - 1, t). \tag{5.12}$$

Let us label

$$M_- = \begin{pmatrix} \frac{-1}{\sqrt{2}} & \frac{1}{\sqrt{2}} \\ 0 & 0 \end{pmatrix} \quad \text{and} \quad M_+ = \begin{pmatrix} 0 & 0 \\ \frac{1}{\sqrt{2}} & \frac{1}{\sqrt{2}} \end{pmatrix}.$$

Thus

$$\Psi(n, t+1) = M_-\Psi(n+1, t) + M_+\Psi(n-1, t). \tag{5.13}$$

Equation (5.13) is a difference equation with $\Psi(0, 0) = \begin{pmatrix} 1 \\ 0 \end{pmatrix}$ and $\Psi(n, 0) = \begin{pmatrix} 0 \\ 0 \end{pmatrix}, \forall\, n \neq$ 0 as initial conditions (Eq. (5.7)).

The purpose of this analysis is to find analytical expressions for $\Psi_L(n, t)$ and $\Psi_R(n, t)$. To do so, we compute the Discrete Time Fourier Transform of Eq. (5.13). The Discrete Time Fourier Transform is given as follows.

Definition 5.1.1. Discrete Time Fourier Transform. *The Discrete Time Fourier Transform is part of the family of Fourier transforms. It transforms a function $f(n)$ of a discrete "time" variable $n \in \mathbb{Z}$ into a continuous, periodic spectrum $F(e^{i\omega})$. Let $f : \mathbb{Z} \to \mathbb{C}$ be a complex function over the integers $\Rightarrow$ its Discrete Time Fourier Transform (DTFT) $\tilde{f} : [-\pi, \pi] \to \mathbb{C}$ is given by*

$$F(e^{i\omega}) = \sum_{n=-\infty}^{\infty} f(n)e^{-in\omega},$$

and its inverse is given by

$$f(n) = \frac{1}{2\pi} \int_{-\pi}^{\pi} F(e^{i\omega})e^{in\omega} d\omega.$$

Ambainis et al. [96] employ the following slight variant of the DTFT:

$$\tilde{f}(k) = \sum_{n} f(n)e^{ik}, \tag{5.14}$$

where $f : \mathbb{Z} \to \mathbb{C}$ and $\tilde{f} : [-\pi, \pi] \to \mathbb{C}$. Corresponding inverse DTFT is given by

$$f(n) = \frac{1}{2\pi} \int_{-\pi}^{\pi} \tilde{f}(k)e^{-ik} dk. \tag{5.15}$$

So, using Eq. (5.14) we have

$$\tilde{\Psi}(k, t) = \sum_{n} \Psi(n, t)e^{ikn}. \tag{5.16}$$

Using Eq. (5.13) we obtain

$$\tilde{\Psi}(k, t+1) = \sum_n (M_- \Psi(n+1, t) + M_+ \Psi(n-1, t)) e^{ikn}. \qquad (5.17)$$

After some algebra we get

$$\tilde{\Psi}(k, t+1) = M_k \tilde{\Psi}(k, t), \quad \text{where } M_k = e^{-ik} M_- + e^{ik} M_+ = \frac{1}{\sqrt{2}} \begin{pmatrix} -e^{-ik} & e^{-ik} \\ e^{ik} & e^{ik} \end{pmatrix}. \qquad (5.18)$$

Thus

$$\tilde{\Psi}(k, t) = \begin{pmatrix} \tilde{\Psi}_L(k, t) \\ \tilde{\Psi}_R(k, t) \end{pmatrix} = M_k^t \tilde{\Psi}(k, 0), \quad \text{where } \tilde{\Psi}(k, 0) = \begin{pmatrix} 1 \\ 0 \end{pmatrix}. \qquad (5.19)$$

Our problem now consists on diagonalizing the (unitary) matrix M_k in order to calculate M_k^t (Theorem 2). If M_k has eigenvalues $\{\lambda_k^1, \lambda_k^2\}$ and eigenvectors $|\Phi_k^1\rangle, |\Phi_k^2\rangle$ then

$$M_k = \lambda_k^1 |\Phi_k^1\rangle\langle|\Phi_k^1| + \lambda_k^2 |\Phi_k^2\rangle\langle\Phi_k^2|. \qquad (5.20)$$

Using Def. 2.1.13 we find

$$M_k^t = (\lambda_k^1)^t |\Phi_k^1\rangle\langle\Phi_k^1| + (\lambda_k^2)^t |\Phi_k^2\rangle\langle\Phi_k^2|. \qquad (5.21)$$

It is shown in [96, 113] that

$$\lambda_k^1 = e^{i\omega_k}, \lambda_k^2 = e^{i(\pi - \omega_k)}, \quad \text{where } \omega_k \in \left[-\frac{\pi}{2}, \frac{\pi}{2}\right] \text{ and } \sin(\omega_k) = \frac{\sin k}{\sqrt{2}} \qquad (5.22)$$

and

$$\Phi_k^1 = \frac{1}{\sqrt{2[(1 + \cos^2(k)) + \cos(k)\sqrt{1 + \cos^2 k}]}} \begin{pmatrix} e^{-ik} \\ \sqrt{2}e^{i\omega_k} + e^{-ik} \end{pmatrix}, \qquad (5.23a)$$

$$\Phi_k^2 = \frac{1}{\sqrt{2[(1 + \cos^2(\pi - k)) + \cos(\pi - k)\sqrt{1 + \cos^2(\pi - k)}]}} \begin{pmatrix} e^{-ik} \\ -\sqrt{2}e^{-i\omega_k} + e^{-ik} \end{pmatrix}. \qquad (5.23b)$$

From Eqs. (5.22), (5.23a), and (5.23b) we compute the Fourier-transformed amplitudes $\tilde{\Psi}_L(n, t)$ and $\tilde{\Psi}_R(n, t)$:

$$\tilde{\Psi}_L(n, t) = \frac{e^{-ik}}{2\sqrt{1 + \cos^2 k}} (e^{i\omega_k t} - (-1)^t e^{-i\omega_k t}), \qquad (5.24a)$$

$$\tilde{\Psi}_R(n, t) = \frac{1}{2}\left(1 + \frac{\cos k}{\sqrt{1 + \cos^2 k}}\right)e^{i\omega_k t} + \frac{(-1)^t}{2}\left(1 - \frac{\cos k}{\sqrt{1 + \cos^2 k}}\right)e^{-i\omega_k t}. \qquad (5.24b)$$

Using Eq. (5.1.1) on Eqs. (5.24a) and (5.24b), we prove the following theorem:

Theorem 1. *Let $|\Psi\rangle_0 = |0\rangle_p \otimes |0\rangle_c$ be the initial state of a discrete quantum walk on an infinite line with coin and shift operators given by Eqs. (5.1) and (5.2), respectively* $\Rightarrow$

$$\Psi_L(n, t) = \frac{1}{2\pi}\int_{-\pi}^{\pi} \frac{-i e^{ik}}{2\sqrt{1 + \cos^2 k}}(e^{-i(\omega_k t - kn)})dk,$$

$$\Psi_R(n, t) = \frac{1}{2\pi}\int_{-\pi}^{\pi}\left(1 + \frac{\cos k}{\sqrt{1 + \cos^2 k}}\right)(e^{-i(\omega_k t - kn)})dk,$$

where $\omega_k = \sin^{-1}\left(\frac{\sin k}{\sqrt{2}}\right)$ and $\omega_k \in \left[\frac{-\pi}{2}, \frac{\pi}{2}\right]$.

The amplitudes for even n (odd n) at odd t (even t) are zero, as can be inferred from the definition of the quantum walk. Now we have an analytical expression for $\Psi_L(n, t)$ and $\Psi_R(n, t)$, and taking into account that $P(n, t) = |\Psi_L(n, t)|^2 + |\Psi_R(n, t)|^2$, we are interested in studying the asymptotical behavior of $\Psi(n, t)$ and $P(n, t)$. Integrals in Theorem 1 are of the form

$$I(\alpha, t) = \frac{1}{2\pi}\int_{-\pi}^{\pi} g(k)e^{i\phi(k, \alpha)t}dk, \quad \text{where } \alpha = n/t \, (= \text{position/number of steps}).$$

The asymptotical properties of this kind of integral can be studied using the method of stationary phase [123, 124], a standard method in complex analysis. Using such a method, the authors of [96, 113] reported the following theorems and conclusions:

Theorem 2. *Let $\epsilon > 0$ be any constant, and α be in the interval $\left(\frac{-1}{\sqrt{2}} + \epsilon, \frac{1}{\sqrt{2}} - \epsilon\right)$. Then, as $t \to \infty$, we have (uniformly in n)*

$$p_L(n, t) \sim \frac{2}{\pi\sqrt{1 - 2\alpha^2}t}\cos^2\left(-\omega t + \frac{\pi}{4} - \rho\right),$$

$$p_R(n, t) \sim \frac{2(1 + \alpha)}{\pi(1 - \alpha)\sqrt{1 - 2\alpha^2}t}\cos^2\left(-\omega t + \frac{\pi}{4}\right),$$

where $\omega = \alpha\rho + \theta$, $\rho = \arg(-B + \sqrt{\Delta})$, $\theta = \arg(B + 2 + \sqrt{\Delta})$, $B = \frac{2\alpha}{1-\alpha}$, and $\Delta = B^2 - 4(B + 1)$.

Theorem 3. *Let $n = \alpha t \to \infty$ with α fixed. In case $\alpha \in (-1, -1/\sqrt{2}) \cup (1/\sqrt{2}, 1) \Rightarrow \exists\, c > 1$ for which $p_L(n, t) = O(c^{-n})$ and $p_R(n, t) = O(c^{-n})$.*

Conclusions

1. **Quasi-uniform behavior.** The wavefunction $\Psi_L(n, t)$ and $\Psi_R(n, t)$ (Theorem 1) is almost uniformly spread over the region for which α is in the interval $[-1/\sqrt{2}, 1/\sqrt{2}]$ (Theorem 2), and shrinks quickly outside this region (Theorem 3). Furthermore, by integrating the probability functions from Theorem 2, it is possible to see that almost all of the probability is concentrated in the interval $[(-1/\sqrt{2} + \epsilon)t, (1/\sqrt{2} - \epsilon)t]$. In fact, the exact probability value in that interval is $P = 1 - \frac{2\epsilon}{\pi} - \frac{O(1)}{t}$.

2. **Standard deviation.** According to [96, 113], the zeroth and second moments of the probability distribution from Theorem 2 are $\mu_1 = \frac{1-\sqrt{2}}{\sqrt{2}}$ and $\mu_2 = \frac{\sqrt{2}-1}{\sqrt{2}}$, respectively. Being rigorous, and taking into account that both moments were computed using *normalized* (over the total number of steps t) probability distributions, then the variance of the Hadamard walk is given by Eq. (4.5):

$$\sigma_{\hat{H}}^2 = \mu_2 - \mu_1^2 = \left[\frac{\sqrt{2}-1}{\sqrt{2}}\right] t - \left(\left[\frac{1-\sqrt{2}}{\sqrt{2}}\right] t\right)^2. \tag{5.25}$$

That is, $\sigma_{\hat{H}}^2 = O(t^2)$ and, consequently,

$$\sigma_{\hat{H}} = O(t). \tag{5.26}$$

However, the second moment has also been interpreted [125, 126] as the actual variance of the probability distribution given in Theorem 2. Furthermore, by introducing a novel method to compute the probability distribution X of the unrestricted DQWL, it was shown in [126] that $\frac{\sigma(X)}{t} \to \sqrt{\frac{\sqrt{2}-1}{2}}$ as $t \to \infty$. In any case, the standard deviation of the unrestricted Hadamard DQWL is $O(t)$ and that result is in contrast with the standard deviation of an unrestricted classical random walk on a line, which is $O(\sqrt{t})$ (Eq. (4.12)).

3. **Mixing time.** It was shown in [96, 113] that an unrestricted Hadamard DQWL has a linear mixing time $\tau_\epsilon^{(q)} = O(t)$, where t is the number of steps. Furthermore, $\tau_\epsilon^{(q)}$ was compared with the corresponding mixing time of a classical random walk on a line, which is quadratic, i.e. $\tau_\epsilon^{(c)} = O(t^2)$.

In order to properly bound and evaluate the impact of this result in the fields of quantum walks and quantum computation, a few clarifications are needed.

(a) The mixing time measure used in this case is not the same as Eq. (4.2.5), the reason being that *unitary* Markov chains in **finite** state space (such as finite graph analogues of quantum walks) have no stationary distribution [96, Sec. 2]. Instead, the mixing time measure proposed is given by

Definition 5.1.2.   Instantaneous Mixing Time. $\tau_\epsilon = \max_u \min_t \{t| \; ||P_u(t) - \pi|| \le \epsilon\}$

which is a more relaxed definition in the sense that it measures the first time that the current probability distribution $P_u(t)$ is ϵ-close to the stationary distribution, *without the requirement of continuing being ϵ-close for all future steps.*

(b) The stationary distribution of an unrestricted classical random walk on a line is the binomial distribution, spread all over $\mathbb{Z}$. The only difference between P_t, the probability distribution of an unrestricted classical random walk on a line at step t, and its limiting distribution P is the numerical value of the probability assigned to each node, as the shape of the distribution is the same. Although the binomial distribution can be *roughly* approximated by a uniform distribution for large values of t, depending on the precision we need for a certain task, that comparison is not precise.

   We can adopt the hitting time of an unrestricted classical random walk on a line together with Theorem 10 to figure out its corresponding mixing time. As shown in our chapter on classical random walks, the hitting time of an unrestricted classical random walk on a line depends on the region we are looking into. Specifically, the hitting time is $O(\sqrt{t})$ for $k \ll t$ and $O(2^t)$ for $k \approx t$ (Eqs. (4.17) and (4.18)). Thus, to hit node k with equal probabilities $P_{t_k} = P_k$ may depend on the region where k is located. For example, it may take $O(\sqrt{t})$ if $k \ll t$ and $O(2^t)$ if $k \approx t$. As previously expressed, it seems that more analysis and new methods for studying mixing times on unrestricted classical random walks are required, particularly within the framework of algorithm development.

   So, comparing mixing times for quantum and classical unrestricted walks on a line is not necessarily clear and straightforward. Furthermore, and in order to reduce complexity in the analysis of algorithms, the infiniteness property of unrestricted classical random walks can sometimes be relaxed and properties of classical random walks on finite lines are used instead [43].

   This is indeed the case in the comparison of mixing times for classical and quantum walks on a line. As shown in Eq. (4.20), the hitting time (and therefore its mixing time) of a classical random walk on a line with reflecting barriers is $O(t^2)$, where t is the number of steps.

Discrete Path Integral Analysis of the Hadamard Walk

A different proposal to study the properties of quantum walks, based on combinatorics and the method given in [112] to quantify quantum state amplitudes, has been delivered in [96, 121, 122]. The main idea behind this approach is to count the number of paths that take a quantum walker from point a to point b. Thus, this approach can also be seen as a discrete path-integral method. Let us begin by stating the following lemma:

Lemma 1. *[96, 112]. Let $t \in [-n, n) \cap \mathbb{Z}$ and $l = \frac{t-n}{2}$. The amplitudes of position n after t steps of the Hadamard walk are*

$$\psi_L(n, t) = \frac{1}{\sqrt{2^t}} \sum_k \binom{l-1}{k}\binom{t-l}{k}(-1)^{l-k-1}, \qquad (5.27a)$$

$$\psi_R(n, t) = \frac{1}{\sqrt{2^t}} \sum_k \binom{l-1}{k-1}\binom{t-l}{k}(-1)^{l-k}. \qquad (5.27b)$$

It was shown in [96] that the probabilities computed from those amplitudes of Lemma 1 can be expressed using Jacobi polynomials. Furthermore, it was shown in [122] that both Schrödinger and combinatorial approaches are equivalent.

Theorem 4. *Let $n \in \mathbb{N} \cup \{0\}$ and $J_v^{(a,b)}(z)$ be the normalized degree v Jacobi polynomial with $J_v^{(a,b)}$ as its constant term. Let us also define $v = \frac{(t-n)}{2} - 1$. Then*

$$P_l(n, t) = 2^{-n-2}(J_v^{(0,n+1)})^2, \qquad (5.28a)$$

$$P_R(n, t) = \left(\frac{t+n}{t-n}\right)^2 2^{-n-2}(J_v^{(1,n)})^2, \qquad (5.28b)$$

with $p_L(-n, t) = p_L(n-2, t)$ and $p_R(-n, t) = \left(\frac{t-n}{t+n}\right)^2 p_R(n, t)$.

A slight variation of this approach is given in [127]. An alternative method based on combinatorics and decompositions of unitary matrices has been proposed in [126, 128–130]. Also, Katori et al. [131] apply group theory to analyze symmetry properties of quantum walks on a line and, along the same line of thought, Chandrashekar et al. [132] have proposed a generalized version of the discrete quantum walk with coins living in *SU(2)*.

Unrestricted DQWL With a General Coin and With Several Coins
The study of the Hadamard walk is relevant to the field of quantum walks not only as an example but also because of the fact that some important properties shown by the Hadamard walk (for example, its standard deviation and mixing time) are shared by any quantum walk on the line.

In [133] it was shown that for a general unbiased initial coin state

$$|\psi(x,0)\rangle = \sqrt{\eta}(|0\rangle_c + e^{i\alpha}\sqrt{1-\eta}|1\rangle_c) \otimes |0\rangle_p \qquad (5.29)$$

and a single step (in Fourier space) of the quantum walk

$$|\tilde{\psi}(k,t+1)\rangle = \tilde{C}_k|\tilde{\psi}(k,t)\rangle,$$

where

$$\tilde{C}_k = \begin{pmatrix} \sqrt{\rho}e^{ik} & \sqrt{1-\rho}e^{i(\theta+k)} \\ \sqrt{1-\rho}e^{i(-k+\phi)} & -\sqrt{\rho}e^{i(-k+\theta+\phi)} \end{pmatrix} \qquad (5.30)$$

is the Fourier-transformed version of the most general two-dimensional coin operator

$$\mathbf{C_2} = \begin{pmatrix} \sqrt{\rho} & \sqrt{1-\rho}e^{i\theta} \\ \sqrt{1-\rho}e^{i\phi} & -\sqrt{\rho}e^{i(\theta+\phi)} \end{pmatrix}$$

with $\theta, \phi \in [0,\pi]$ and $\rho \in [0,1]$, we can express a t-step quantum walk on a line as

$$|\tilde{\psi}(k,t+1)\rangle = \tilde{C}_k^t|\tilde{\psi}(k,0)\rangle, \text{ where } |\tilde{\psi}(k,0)\rangle = \begin{pmatrix} \sqrt{\eta} \\ e^{i\alpha}\sqrt{1-\eta} \end{pmatrix} \otimes |k\rangle. \qquad (5.31)$$

If $\tilde{C}_k$ is expressed in terms of its eigenvalues $\lambda_k^\pm$ and eigenvectors $|\lambda_k^\pm\rangle$ then $\tilde{C}_k^t = (\lambda_k^+)^t|\lambda_k^+\rangle\langle\lambda_k^+| + (\lambda_k^-)^t|\lambda_k^-\rangle\langle\lambda_k^-|$, and Eq. (5.31) can be written as

$$|\tilde{\psi}(k,t+1)\rangle = (\lambda_k^+)^t|\lambda_k^+\rangle\langle\lambda_k^+|\tilde{\psi}(k,0)\rangle + (\lambda_k^-)^t|\lambda_k^-\rangle\langle\lambda_k^-|\tilde{\psi}(k,0)\rangle \qquad (5.32)$$

with

$$(\lambda_k^\pm)^t\langle\lambda_k^\pm|\tilde{\psi}(k,0)\rangle = \frac{(\lambda_k^\pm)^t}{n_k^\pm}e^{-ik}\left[\sqrt{\eta} - \sqrt{\frac{1-\eta}{1-\rho}}e^{i(\theta+\alpha)}(\sqrt{\rho} \mp e^{i(k-\delta)}e^{\mp i\omega_k})\right], \qquad (5.33)$$

where $\delta = (\theta+\phi)/2$, $\sin(\omega_k) = \sqrt{\rho}\sin(k-\delta)$, $\lambda_k^\pm = \pm e^{i\delta}e^{\pm i\omega_k}$, $n_k = \sqrt{\frac{2[1\mp\sqrt{\rho}\cos(k-\delta\mp\omega_k)]}{1-\rho}}$, $\lambda^\pm = \pm e^{i\delta}e^{\pm i\omega_k}$, and

$$|\lambda^\pm\rangle = \frac{1}{n_k^\pm}\begin{pmatrix} e^{ik} \\ e^{i\theta}(\lambda^\pm - \sqrt{\rho}e^{ik})/\sqrt{1-\rho} \end{pmatrix}.$$

As in the Hadamard walk case, the properties of the quantum walk defined by Eqs. (5.33) and (5.31) may be studied by inverting the Fourier transform and using methods of complex analysis. Let us concentrate on the phase factors $\alpha \in \mathbb{R}$ of the coin initial state (Eq. (5.29)) and $\theta \in \mathbb{R}$ of the coin operator (Eq. (5.30)). Note that we can choose many pairs of values (α, θ) for any phase factor $r = \alpha + \theta$. So, if we fix a value for θ (i.e. if we use only one coin operator)

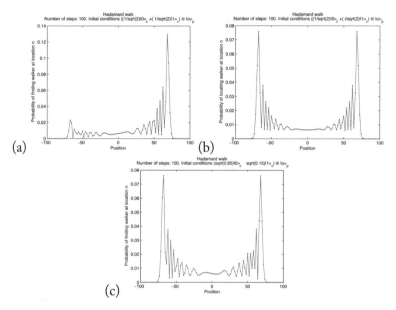

FIGURE 5.2: Graph (a) was computed using coin initial state $|\psi\rangle_0 = |0\rangle_c \otimes |0\rangle_p$. Graphs (b) and (c) had $|\psi\rangle = \frac{1}{\sqrt{2}}(|0\rangle_c + i|1\rangle_c) \otimes |0\rangle_p$ and $|\psi\rangle = \sqrt{0.85}|0\rangle_c - \sqrt{0.15}|1\rangle_c) \otimes |0\rangle_p$ as coin initial states, respectively. Note that symmetry in the probability distribution can be achieved by using coin initial states with either complex or real relative phase factors [133]. All graphs were computed from 100-step Hadamard quantum walks on a line with Eq. (5.2) as shift operator.

we can always vary the initial coin state $|\psi(x, 0)\rangle$ (Eq. (5.29)) to get a value for α so that we can compute a quantum walk with a certain phase factor value r. It is in this sense we say that the study of a Hadamard walk suffices to analyze the properties of all unrestricted quantum walks on a line. In Fig. 5.2 we show the probability distributions of three Hadamard walks with different initial coin states.

The effect of different and multiple coins has been studied by several authors. In [134, 135], Konno and Inui have examined probability distributions computed with quantum walks on a line using three- and four-dimensional coins, respectively. The results shown in [134] have some similarities with the quantum walks with maximally entangled coins reported in [136] in the sense that both quantum walks tend to concentrate most of their probability distributions about the origin of the walk. Additionally, Ribeiro et al. [137] have considered quantum walks with several biased coins applied aperiodically, D'Alessandro et al. [138] have studied non-stationary quantum walks on a cycle using different coin operators at each computational step, and Feinsilver and Kocik [139] have proposed the use of Krawtchouk matrices (via tensor powers of the Hadamard matrix) for calculating quantum amplitudes. In [127], Brun et al. analyzed the behavior of a quantum walk on the line using both M two-dimensional coins and single coins of 2^M dimension, and Bañulus et al. [140] have studied the behavior of quantum walks with a time-dependent coin.

Ermann et al. [141] have inspected the decoherence of quantum walks with a complex coin, where the coin is part of a larger quantum system and Chandrashekar et al. [142] have studied symmetries and noise effects on coined discrete quantum walks. Finally, Kendon et al. [125, 143, 144] have extensively studied the computational consequences of coin decoherence (i.e. interaction with the environment) in quantum walks.

5.1.3 Discrete Quantum Walk With Boundaries

The properties of discrete quantum walks on a line with one and two absorbing barriers were first studied in [96]. For the semi-infinite discrete quantum walk on a line, Theorem 5 was reported.

Theorem 5. *Let us denote by p_∞ the probability that the measurement of whether the particle is at the location of the absorbing boundary (location 0 in [96]) $\Rightarrow p_\infty = \frac{2}{\pi}$.*

Theorem 5 is in stark contrast with its classical counterpart (Eq. (4.15)), as the probability of eventually being absorbed is equal to unity.

The case of a quantum walk on a line with two absorbing boundaries was also studied in [96], and their main result is given in Theorem 6.

Theorem 6. *For each $n > 1$, let p_n be the probability that the process eventually exits to the left. Also define q_n to be the probability that the process exits to the right. Then*

$$\text{(i) } \forall\, n > 1 \;\Rightarrow\; p_n + q_n = 1,$$

$$\text{(ii) } \lim_{n \to \infty} p_n = \frac{1}{\sqrt{2}}.$$

Theorems 5 and 6 are revisited in [145] with detailed corresponding proofs using both Fourier transform and path counting approaches. Also, [145] proves some conjectures given in [146]. Finally, Konno studied the properties of quantum walks with boundaries using a set of matrices derived from a general unitary matrix together with a path counting method [103, 147].

5.2   QUANTUM WALKS ON GRAPHS

Classical random walks on graphs have been crucial to the development of stochastic algorithms [23]. In consequence, quantum walks on graphs has become an active area of research in quantum computation. A gentle introduction to the main ideas about discrete and continuous quantum walks on graphs, as well as to the quantification of resources required for their implementation, is given in [148]. Also, [149] presents numerical simulations of quantum walks in higher dimensions using separable and non-separable coin operators.

In [150], Aharonov et al. studied several properties of quantum walks on undirected graphs. Motivated by the importance of stationary distributions of Markov Chains (Theorem 8), the quantum counterpart of a stationary distribution is studied in [150]. Their first finding consisted in proving that, if we adopt the classical definition of stationary distribution (Def. 4.1.11), then quantum walks do not converge to any stationary state nor to any stationary distribution.

In order to review the contributions of [150] and other authors, let us begin by formally introducing the following elements. Let $G = (V, E)$ be a d-regular graph (Def. 4.2.1) with $|V| = n$. Note that graphs studied in this section are *finite*, as opposed to the unrestricted line we used in the beginning of this chapter. Let $\mathcal{H}_v$ be the Hilbert space spanned by states $|v\rangle$ where $v \in V$. Also, we define $\mathcal{H}_A$, the coin space, as an auxiliary Hilbert space of dimension d spanned by the basis states $\{|i\rangle | i \in \{1, \ldots, d\}\}$, and $\hat{C}$, the coin operator, as a unitary transformation in $\mathcal{H}_A$. Finally, label each directed edge with a number between 1 and d so that the directed edges form a permutation (for Cayley graphs the labeling of a directed edge is simply the generator associated with the edge). Now we define a shift operator $\hat{S}$ on $\mathcal{H}_v \otimes \mathcal{H}_A$ such that $\hat{S}|a, v\rangle = |a, u\rangle$, where u is the ath neighbor of v (since edge labeling is a permutation then $\hat{S}$ is unitary). Finally, we define one step of the quantum walk on G as $\hat{U} = \hat{S}(\hat{C} \otimes \hat{I})$.

As in the study of quantum walks on a line, if $|\psi\rangle_0$ is the quantum walk initial state then a quantum walk on a graph G can be defined as

$$|\psi\rangle_t = \hat{U}^t |\psi\rangle_0. \tag{5.34}$$

Before introducing the concept of quantum limiting distribution, we provide an example of a quantum walk on a graph: a discrete quantum walk on a cycle.

Example. Discrete quantum walk on a cycle. Let G_{cyc} be a cycle with n nodes (see Fig. 5.3). A quantum walk on G_c acts on a total Hilbert space $\mathcal{H}^2 \otimes \mathcal{H}^n$. For the sake of this example, we employ the Hadamard coin operator given by Eq. (2.4) and the shift operator defined by $\hat{S}|0, j\rangle = |0, j+1 \mod n\rangle$ and $\hat{S}|1, j\rangle = |0, j-1 \mod n\rangle$.

Now we discuss the definition and properties of limiting distributions for quantum walks on graphs. Suppose we begin a quantum walk with initial state $|\psi\rangle_0$. Then, after t steps, the probability distribution of the graph nodes induced by Eq. (5.34) is given by

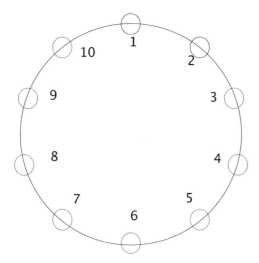

FIGURE 5.3: Quantum walk on a cycle. A cycle is a 2-regular graph which can be viewed as a Cayley graph of the group $\mathbb{Z}$ with generators $1, -1$. The cycle shown in this figure has ten vertices.

Definition 5.2.1.   Probability distribution on the nodes of G. *Let v be a node of G and $\mathcal{H}^d$ be the coin Hilbert space. Then*

$$P_t(v|\psi_0) = \sum_{i \in \{1,\dots,d\}} |\langle i, v | \psi \rangle_t|^2.$$

If probability distributions P_0, P_1 at time 0 and 1 are different, it can be proved [150] that P_t does not converge. However, if we compute the *average* of distributions over time

Definition 5.2.2.   Averaged probability distribution.

$$\bar{P}_t(v|\psi_0) = \frac{1}{T} \sum_{t=0}^{T-1} P_t(v|\psi_0),$$

we can obtain the following result.

Theorem 7. *[150]. Let $|k\rangle$, λ_k denote the eigenvectors and corresponding eigenvalues of $\hat{U}$. Then, for an initial state $|\psi\rangle_0 = \sum_k a_k |k\rangle$*

$$\lim_{t \to \infty} \bar{P}_t(v|\psi_0) = \sum_{i,j,a} a_i a_j^* \langle a, v|i\rangle \langle j|a, v\rangle,$$

where the sum is only on pairs i, j such that $\lambda_i = \lambda_j$.

If all the eigenvalues of $\hat{U}$ are distinct, the limiting distribution takes a simple form. Let $p_i(v) = \sum_{i \in \{1,...,d\}} |\langle i, v|k \rangle|^2$, i.e. $p_i(v)$ is the probability to measure node v in the eigenstate $|k\rangle$. Then it is possible to prove that for an initial state $|\psi\rangle_0 = \sum_k a_k |k\rangle \Rightarrow \lim_{T \to \infty} \bar{P}_t(v|\psi_0) = \sum_i |a_i|^2 p_i(v)$ [150]. Using this fact it is possible to prove the following theorem.

Theorem 8. [150] *Let $\hat{U}$ be a coined quantum walk on the Cayley graph of an Abelian group, such that all eigenvalues of $\hat{U}$ are distinct. Then the limiting distribution π (Def. 5.2.2) is uniform over the nodes of the graph, independent of the initial state $|\psi\rangle_0$.*

Using Theorem 8 we compute the limiting distribution of a quantum walk on a cycle.

Theorem 9. *The limiting distribution π for the coined quantum walk on the n-cycle, with n odd, and with the Hadamard operator as coin, is uniform on the nodes, independent of the initial state $|\psi\rangle_0$.*

Several other important results for quantum walks on a graph are delivered in [150]. Among them, we mention some results on mixing times.

Definition 5.2.3. **Average Mixing time.** *The mixing time M_ϵ of a quantum Markov chain with initial state $|k, v\rangle$ is given by*

$$M_\epsilon = \min\{T | \forall t \geq T \Rightarrow ||\bar{P}_t(k, v) - \pi(k, v)|| \leq \epsilon\}.$$

Theorem 10. *For the quantum walk on the n-cycle, with n odd, and the Hadamard operator as coin, we have*

$$M_\epsilon \leq O\left(\frac{n \log n}{\epsilon^3}\right).$$

So, the mixing time of a quantum walk on a cycle is $O(n \log n)$. The mixing time of corresponding classical random walk on a circle is $O(n^2)$ (Eq. (4.22)). Now we focus on a general property of mixing times.

Theorem 11. *For a general quantum walk on a bounded degree graph, the mixing time is at most quadratically faster than the mixing time of the simple classical random walk on that graph.*

The properties of the wavefunction of a quantum particle randomly walking on a circle have been studied in [151], and some details of limiting distributions of quantum walks on cycles are shown in [152] as well as in [153]. Also, the effect of using different coins on the behavior of quantum walks on an *n*-cycle as well as in graphs of higher degree has been studied in [133]. Finally, a standard deviation measure for quantum walks on circles is introduced in [154].

Another graph studied in quantum walks is the hypercube, defined by the following.

Definition 5.2.4. The hypercube. *The hypercube is an undirected graph with 2^n nodes, each of which is labeled by a binary string of n bits. Two nodes $\vec{x}, \vec{y}$ in the hypercube are connected by an edge if $\vec{x}, \vec{y}$ differ only by a single bit flip, i.e. if $|\vec{x} - \vec{y}| = 1$, where $|\vec{x} - \vec{y}|$ is the Hamming distance between $\vec{x}$ and $\vec{y}$. As an example, the three-dimensional hypercube is shown in Fig. 6.1.*

In [155], Moore and Russell derived values for *the two notions* of mixing times we have studied (Defs. 5.1.2 and 5.2.3) for continuous and discrete quantum walks on the hypercube. As for the discrete quantum walk, [155] begins by defining Grover's operator as coin operator.

Definition 5.2.5. Grover's operator. *Let $\mathcal{H}$ be an n-dimensional Hilbert space and $|i\rangle$ be the canonical basis for $\mathcal{H}$ and $|\psi\rangle = \frac{1}{\sqrt{n}}\sum_{i=0}^{n-1}|i\rangle$. Then we define Grover's operator as $\hat{G} = |\psi\rangle\langle\psi| - \hat{I}$.*

Additionally, their shift operator is given by

$$\hat{S} = \sum_{d=0}^{n-1}\sum_{\vec{x}}|d, \vec{x} \oplus \vec{e_d}\rangle\langle d, \vec{x}| \tag{5.35}$$

where $\vec{e_d}$ is the ith basis vector of the n-dimensional hypercube. So, the quantum walk on the hypercube proposed in [155] can be written as

$$|\psi\rangle_t = \hat{U}^t|\psi\rangle_0 = [\hat{S}(\hat{G} \otimes \hat{I}_n)]^t|\psi\rangle_0 \tag{5.36}$$

for a given initial state $|\psi\rangle_0$. Using a Fourier transform approach as in [113], it was proved in [155] that

Theorem 12. *For the discrete quantum walk defined in Eq. (5.36), its instantaneous mixing time (Def. 5.1.2) is given by $t = \frac{k\pi}{4}n$, i.e. $t = O(n)$, with $\epsilon = O(n^{-7/6})$ for all odd k.*

Reference [155] has several other contributions, and among those we would like to briefly mention that its authors elaborate on the fact that the relationship between different definitions of mixing times (i.e. instantaneous and average mixing times) for continuous and discrete quantum walks is not clear. Additionally, [155] provides analytical expressions for eigenvalues and corresponding eigenvectors of the evolution operator defined in Eq. (5.36) which were later used in [156] for the design of a search algorithm based on a discrete quantum walk (more on this in the following section).

According to Theorem 11, the speedup that can be provided by a quantum walk on a graph is not enough to exponentially outperform classical walks. So, other parameters of quantum walks have been investigated, among them their *hitting time*. In [157], Kempe offers an analysis of hitting time of discrete quantum walks on the hypercube. Due to the potential service of hitting times in the construction of quantum algorithms, we shall analyze [157] in detail in Chapter 6 (algorithms based on quantum walks).

The sub-field of quantum walks on graphs is wide and rich. As a result, there are several interesting works which we have not covered in detail in this lecture due to space restrictions. However, we would like to point out relevant works to the interested reader. We start by mentioning the numerical simulations of quantum walks on graphs shown in [133], particularly the "localization" phenomenon due to the use of Grover's operator (Eq. (5.2.5)) in a two-dimensional quantum walk. Inspired by this phenomenon, Innui et al. proved in [158] that the key factor behind this localization phenomenon is the degeneration of the eigenvectors of corresponding evolution operator. In [159], Gottlieb et al. studied the convergence of coined quantum walks in $\mathbb{R}^d$. In [160], Feldman and Hillery have studied the relationship between quantum walks on graphs and scattering theory. Also, López-Acevedo and Gobron [161] delivered an algebraic oriented analysis of quantum walks on Cayley graphs, Montanaro presented in [162] a study on quantum walks on directed graphs, Krovi and Brun [163] have studied quantum walks (and their hitting times) on quotient graphs as well as links between those quantum walks and the group theory properties of Cayley graphs (for an extended work on this last topic, see [164]).

5.3 MORE CONSIDERATIONS ON CLASSICAL AND QUANTUM WALKS

The links between classical and quantum versions of random walks have been studied by several authors under different perspectives:

(1) Some authors, among them Watrous [165], have been interested in simulating classical random walks using quantum walks. Studies on this area would provide us not only with interesting computational properties of both types of walks, but also with a deeper insight of the correspondences between the laws that govern computational processes in classical and quantum physical systems.

(2) Some other authors have studied the properties and conditions of transitions from quantum walks into classical random walks. This area of research is interesting not only for exploring computational properties of both kinds of walks, but also because we would provide quantum computer builders (i.e. experimental physicists and engineers) with some criteria and thresholds for testing the quantumness of a quantum computer. Moreover, these studies have allowed the scientific community to reflect on the quantum nature of quantum walks and some of their

implications in algorithm development (in fact, we shall discuss the quantum nature of quantum walks in subsection 5.3.1).

For example, it was shown in [166] that the quantum–classical walk transition could be achieved via two possible methods, in addition to the obvious procedure of performing measurements: decoherence in the quantum coin and the use of higher dimensional coins. Moreover, by using a discrete path approach, it was shown in [130] that introducing a random selection of coins (that is, amplitude components for coin operators are chosen randomly, being under the unitarity constraint) makes quantum walks behave classically. In [167], the authors make use of a family of graphs (e.g. Fig. 6.2(a)) to exemplify the different behavior of (continuous) quantum walks and classical random walks. Then, several authors have addressed the physical and computational properties of decoherence (i.e. interaction with the environment) in quantum walks: Ermann et al. [141] have inspected the decoherence of quantum walks with a complex coin, where the coin is part of a larger quantum system, and Chandrashekar et al. [142] have studied symmetries and noise effects on coined discrete quantum walks. Then, Kendon et al. [125, 143, 144] have extensively studied the computational consequences of coin decoherence in quantum walks. Finally, Alagic and Russell [168] have studied the effects of independent measurements on a quantum walker traveling along the hypercube, and Košík et al. [169] have studied the quantum to classical transition of a quantum walk by introducing random phase shifts in the coin particle.

5.3.1   Are Quantum Walks Really Quantum?

The results presented in this chapter show that superposition and, consequently, interference play an important role in the structure and properties of discrete quantum walks. However, interference is also a characteristic of classical physical systems, like electromagnetic waves. Thus, it makes sense to scrutinize whether the statistical and computational properties of quantum walks are really due to their quantum nature or not.

Arguments in favor of the plausibility of using classical physics for building experiments, which replicate some interference and statistical properties of quantum walks, are given in [170–173], where it was shown that it is possible to develop implementations of a quantum walk on a line purely described by classical physics (wave interference of electromagnetic fields) and still be able to reproduce the variance enhancement that characterizes a discrete quantum walk. For example, the implementation proposed in [172] utilizes the frequency of a light field as walker and the spatial path or the polarization state of the same light field as the coin.

Arguments in favor of the quantum-mechanical nature of quantum walks have been provided by, among others, Kendon and Sanders [174] who showed it would still be necessary to have a quantum-mechanical description of such an implementation in order to account for two properties of a quantum walk: (i) the indivisibility of the quantum walker and (ii) complementarity, which in quantum computation jargon may be stated as follows: *the trade-off between interference and information about the path followed by the walker (knowing the path followed by a quantum particle decreases the sharpness of the interference pattern [111, 175]).* Furthermore, Romanelli et al. [176] showed that the evolution equation of a quantum walk on a line can be separated into two parts: Markovian and interference terms, and that the quadratic increase in the variance of the quantum walker is a consequence of quantum evolution.

Thus it seems that if we are only interested in some statistical properties of quantum walks, like its variance enhancement with respect to classical random walks, we could do with either classical or quantum experimental setups. However, the quantum-mechanical nature of walkers and/or coins is essential in the following cases:

1. From a purely physical point of view, if one is interested in using quantum walks for testing the quantumness of a quantum computer realization, complementarity would be a very helpful resource as it is a property of quantum-mechanical systems that cannot be reproduced in a classical experiment. A similar argument would be applied in the case of using complementary as a computational resource.

2. Quantum entanglement has been incorporated into quantum walks research either as a result of performing a quantum walk ([110, 177–179], and [180, Ch. 7]) or as a resource to build new kinds of quantum walks ([136, 181], and [180, Chs. 6 and 7]). Since entanglement is a key component in quantum computation, it is worth keeping in mind that quantum walks can be used either as entanglement generators or as computational processes taking advantage of this quantum-mechanical property. Quantum entanglement is produced in quantum walks due to the use of non-local operators operating on two or more qubits. Examples of non-local operators employed in quantum walks can be found in [180, Chs. 6 and 7].

5.4   CONTINUOUS QUANTUM WALKS

In this section we shall define a continuous quantum walk so that we can use it in Section 5.6, where we present recent advances about the mathematical bonds between discrete and continuous quantum walks. We shall revisit continuous quantum walks in Chapter 6, where we explore how this kind of quantum process is utilized in algorithm development.

In [167], Childs et al. present the following formulation of a continuous classical random walk:

Definition 5.4.1. *Let $G = (V, E)$ be a graph with $|V| = n$ then a continuous time random walk on G can be described by the order n infinitesimal generator matrix M given by*

$$M_{ab} = \begin{cases} -\gamma, & a \neq b, (a, b) \in G \\ 0, & a \neq b, (a, b) \notin G \\ k\gamma, & a = b \text{ and } k \text{ is the valence of vertex } a. \end{cases} \tag{5.37}$$

Following [167, 182], the probability of being at vertex a at time t is given by

$$\frac{\mathrm{d}p_a(t)}{\mathrm{d}t} = -\sum_b M_{ab} p_b(t). \tag{5.38}$$

Now, let us define a Hamiltonian [167, 182] that closely follows Eq. (5.37).

Definition 5.4.2. *Let $\hat{H}$ be a Hamiltonian with matrix elements given by*

$$\langle a|H|b \rangle = \begin{cases} -\gamma, & a \neq b, (a, b) \in G \\ k\gamma, & a = b \text{ where the valence of } a \text{ is } k \\ 0, & \text{otherwise.} \end{cases} \tag{5.39}$$

We can then employ Hamiltonian $\hat{H}$ as given in Eq. (5.39), defined in a Hilbert space $\mathcal{H}$ with basis $\{|1\rangle, |2\rangle, \ldots, |n\rangle\}$, for constructing the Schrödinger equation of a quantum state $|\psi\rangle \in \mathcal{H}$

$$\mathrm{i}\frac{\mathrm{d}\langle a|\psi(t)\rangle}{\mathrm{d}t} = -\sum_b \langle a|H|b\rangle\langle b|\psi(t)\rangle. \tag{5.40}$$

Finally, taking Eqs. (5.39) and (5.40) the unitary operator $\hat{U}$,

$$\hat{U} = \exp(-\mathrm{i}\hat{H}t), \tag{5.41}$$

defines a **continuous quantum walk** on graph G. Note that the continuous quantum walk given by Eq. (5.41) defines a process on continuous time and discrete space.

There is an increasing number of publications on continuous time quantum walks. We would refer the interested reader to the works of Konno on continuous time quantum walks on ultrametric spaces [183] and continuous quantum walks on trees in quantum probability theory [184], de Falco et al. on speed and entropy of continuous quantum walks [185], Mülken et al. on quantum transport on small-world networks [186], and an investigation on continuous time quantum walks by using the Krylov subspace-Lanczos algorithm [187].

5.5 WHETHER DISCRETE OR CONTINUOUS: IS IT QUANTUM RANDOM WALKS OR JUST QUANTUM WALKS?

Chance is an inherent component of every single step of a classical random walk. In other words, there is no way to predict step s_i of a classical random walk, no matter how much information we have about previous steps $s_{i-1}, s_{i-2}, \ldots, s_1, s_0$. We can only tell the probability associated with each possible step s_{i+1}^{j}.

On the other hand, if we carefully analyze Eqs. (2.9) and (2.10) as well as all the evolution equations presented in this chapter, we shall convince ourselves of the fact that quantum evolution is deterministic, i.e. for each computational step denoted by $|\psi\rangle_i$ we can always tell the exact description of step $|\psi\rangle_{i+1}$, as $|\psi\rangle_{i+1} = \hat{U}|\psi\rangle_i$.

So, what is random about a quantum walk? Why are quantum walks candidates for developing quantum counterparts of stochastic algorithms? The answer is: randomness comes from the measurement processes that have to be performed on either the quantum walker(s) or the quantum coin(s). So, the probabilistic nature of quantum measurement allows us to introduce randomness into a quantum-walk-based algorithm. Moreover, we are not restricted to introducing chance only at the end of the quantum algorithm execution but we can also exploit several measurement strategies in order to manipulate quantum systems and produce probability distributions suitable for their use in advantageous algorithms; for example, see the "top hat" probability distribution [111], a quasi-uniform distribution created by running a discrete quantum walk and performing measurements on its constituent elements (or, alternatively, allowing such constituent particles to have some interaction with the environment).

5.6 HOW ARE CONTINUOUS AND DISCRETE QUANTUM WALKS CONNECTED?

The mathematical models of discrete and continuous quantum walks studied in the previous sections present a serious problem: it is not clear at all how to transform discrete quantum walks into continuous quantum walks and vice versa. This is an important issue for two reasons: (1) in the classical case, discrete and continuous random walks are connected via a limit process and (2) it is not natural to have two different kinds of quantum diffusion, one of them with an extra particle (the quantum coin).

In [108], Strauch presents a connection between discrete and continuous quantum walks. He starts by using a simplification [167] of the continuous quantum walk defined by Eq. (5.40), namely,

$$\hat{H}|j\rangle = -\gamma(|j-1\rangle - 2|j\rangle + |j+1\rangle) \tag{5.42}$$

which in [108] is rewritten as

$$i\partial_t \psi(n, t) = -\gamma(\psi(n + 1, t) - 2\psi(n, t) + \psi(n - 1, t)) \tag{5.43}$$

where $\psi(n, t)$ is a complex amplitude at the continuous time t and the discrete lattice position n. Then, [108] uses results from [106, 112] to build a discrete quantum walk represented by the following unitary mapping:

$$\psi_R(n, \tau + 1) = \cos\theta\,\psi_R(n - 1, \tau) - i\sin\theta\,\psi_L(n - 1, \tau), \tag{5.44a}$$

$$\psi_L(n, \tau + 1) = \cos\theta\,\psi_L(n + 1, \tau) - i\sin\theta\,\psi_L(n + 1, \tau), \tag{5.44b}$$

where $\psi_R(n, \tau)$ and $\psi_L(n, \tau)$ are complex amplitudes at the discrete time τ and discrete lattice position n.

Strauch's result focuses on building a unitary transformation $\hat{U} = \exp(-i\hat{\mathbf{H}}t)$ that allows us to transform Eqs. (5.44a) and (5.44b) into Eq. (5.42). There are several important conclusions from the developments shown in [108]:

1. It is indeed possible to transform a discrete quantum walk into a continuous one by means of a limit process (although this is not a straightforward derivation).

2. Strauch's derivation does not use any coin degree. Thus [108] agrees, from a new perspective, with Patel et al. [107] with respect to the irrelevance of the coin degree of freedom in order to obtain the statistical enhancements ($\sigma^2 = O(n)$) that discrete quantum walks show.

CHAPTER 6

Computer Science and Quantum Walks

A key activity in quantum computation is the development of quantum algorithms for solving both classical and quantum problems (this includes simulation of quantum systems). Since classical random walks have been used to develop successful stochastic algorithms, there has been a huge interest in understanding the computational properties of quantum walks over the last few years.

A general strategy for building an algorithm based on quantum walks includes choosing (1) the unitary operators (Eq. (2.9)), for discrete quantum walks, or the Hamiltonians (Eq. (2.10)), for continuous quantum walks, that will be employed to determine the time evolution of the quantum hardware and (2) the measurement operators that will be employed to find out the position of the walker (Eq. (2.11)).

The quantum programmer must bear in mind that the choice of evolution and measurement operators, as well as initial quantum states, will determine the shape and other properties of the resulting probability distribution for the quantum walker. Moreover, a computer scientist interested in algorithms based on quantum walks must keep in mind that making copies of arbitrary quantum states is not possible in general due to the no-cloning theorem (subsection 2.2.3) thus copying variable content is not allowed in principle. Indeed, it is possible to use cloning machines for imperfect quantum state copying, but it would frequently translate into computational and estimation errors. Since any non-reversible gate can be converted into a reversible gate [3, 78, 188], errors due to imperfect quantum state cloning are unnecessary and consequently must be avoided. Employing classical computer simulators of quantum walks [189, 190] can be a fruitful exercise in order to figure out the operators and initial states required for algorithmic applications of quantum walks.

Quantum algorithms based on either discrete or continuous quantum walks are built upon detailed and complex mathematical structures and it is not possible to cover all details in a single chapter. Therefore, we shall devote this chapter to review the fundamental links between quantum walks and computer science (mainly algorithms) and we strongly recommend the reader to go to the corresponding references for more details.

This chapter begins with the definition of an oracle, a key element in both classical and quantum algorithms. We then proceed to examine several algorithms based on discrete quantum walks, including an application for simulated annealing. We then provide a summary of algorithmic results based on continuous quantum walks, which include a novel application of continuous quantum walks in quantum chemistry, and finish this chapter by reviewing recent results about the computational universality of quantum walks.

6.1   ALGORITHMIC APPLICATIONS OF QUANTUM WALKS

Definition 6.1.1.   Oracle. *An oracle is an abstract machine used to study decision problems. It can be thought of as a black box which is able to decide certain decision problems in a single step, i.e. an oracle has the ability to* recognize *solutions to certain problems.*

Oracles are widely used in classical algorithm design. In the context of quantum computation, we also use oracles to *recognize* solutions for the search problem. Additionally, we assume that if an oracle recognizes a solution $|\phi\rangle$ then that oracle is also capable of computing a function with $|\phi\rangle$ as argument.

We are interested in searching for M elements in a space of N elements. To do so, we use an index $x \in S$, where $S = \{0, 1, \ldots, N-1\}$, to enumerate those elements. We also suppose we have a function $f : S \to \{0, 1\}$ such that $f(x) = 1$ if and only if x is one of the elements we are looking for. Otherwise, $f(x) = 0$. An oracle can be written as a unitary operator $\hat{O}$ defined by

$$\hat{O}(|x\rangle|q\rangle) = |x\rangle|q \oplus f(x)\rangle, \tag{6.1}$$

where $|x\rangle$ is the index register, $\oplus$ is addition modulo 2 (the XOR operation in computer science parlance), and the oracle qubit $|q\rangle$ is a single qubit which is flipped if $f(x) = 1$ and is left unchanged otherwise. As shown in [3], we can check whether x is a solution to our search problem by preparing $|x\rangle$, applying the oracle, and checking whether the oracle qubit has been flipped to $|1\rangle$. Grover's algorithm [18], as well as several algorithms we shall review in this chapter, make use of an oracle. A comparison of quantum oracles can be found in [191].

We now proceed to review quantum algorithms based on discrete quantum walks.

6.1.1   Algorithms Based on Discrete Quantum Walks

Let us start by introducing the following problem:

Definition 6.1.2.   Searching in an unordered list. *Suppose we have an unordered list of N items labeled $x_1, x_2, \ldots, x_N$. We want to find one of those elements, say x_i.*

Any classical algorithm would take $O(N)$ steps at least to solve the problem given in Def. 6.1.2. However, one of the jewels of quantum computation, Grover's search algorithm [18], would do much better. By using an oracle and a technique called **Amplitude Amplification**, the search algorithm proposed in [18] would only take $O(\sqrt{N})$ time steps to solve the same search problem. In addition to its intrinsic value for outperforming classical algorithms, Grover's algorithm has relevant applications in computer science, including solutions to the 3-SAT problem (Def. 3.4.7) [109].

In [156], Shenvi et al. proposed an algorithm based on a discrete quantum walk to solve the search problem given in Def. 6.1.2. For the sake of completeness and in order to present the results contained in [156], let us remember the definition of a hypercube.

Definition 6.1.3.   The hypercube. *The hypercube is an undirected graph with 2^n nodes, each of which is labeled by a binary string of n bits. Two nodes $\vec{x}, \vec{y}$ in the hypercube are connected by an edge if $\vec{x}, \vec{y}$ differ only by a single bit flip, i.e. if $|\vec{x} - \vec{y}| = 1$, where $|\vec{x} - \vec{y}|$ is the Hamming distance between $\vec{x}$ and $\vec{y}$. As an example, the three-dimensional hypercube is shown in Fig. 6.1.*

An example of a three-dimensional hypercube can be seen in Fig. 6.1. Since each node of the hypercube has degree n and there are 2^n distinct nodes then the Hilbert space upon which the discrete quantum walk is defined is $\mathcal{H} = \mathcal{H}^n \otimes \mathcal{H}^{2^n}$, and each state $|\psi\rangle \in \mathcal{H}$ is described by a bit string $\vec{x}$ and a direction d. We now define the following coin and shift operators

$$\hat{C} = \hat{C}_0 \otimes \hat{\mathbb{I}} = (-\hat{\mathbb{I}} + 2|s^c\rangle\langle s^c|) \otimes \hat{\mathbb{I}}, \tag{6.2}$$

where $|s^c\rangle$ is the equal superposition over all n directions, i.e. $|s^c\rangle = \frac{1}{\sqrt{n}} \sum_{d=1}^{n} |d\rangle$, and

$$\hat{S} = \sum_{d=0}^{n-1} \sum_{\vec{x}} |d, \vec{x} \otimes \vec{e}_d\rangle\langle d, \vec{x}|, \tag{6.3}$$

where $|\vec{e}_d\rangle$ is the dth basis vector of the hypercube. Using the eigenvalues and eigenvectors of the evolution operator $\hat{U} = \hat{S}\hat{C}$ of the quantum walk on the hypercube [155] in order to build a slightly modified coin operator C' (which works within the algorithm structure as an oracle, Def. 6.1.1) and an evolution operator $\hat{U}'$, and by collapsing the hypercube into a line, the quantum walk designed by evolution operator $\hat{U}'$ is used to search for element $x_{\text{target}} \in \{0, 1\}^n$.

It is claimed in [156] that, after applying $\hat{U}'$ a number of $t_f = \frac{\pi}{2}\sqrt{2^n} = O(\sqrt{N})$ times, the outcome of their algorithm is x_{target} with probability $\frac{1}{2} - O(\frac{1}{n})$. A summary of similarities and differences between this quantum walk algorithm and Grover's algorithm can be found in the last few pages of [156]. Also, Gábris et al. [192] studied the impact of noise on the algorithmic performance given in [156], using a scattering quantum walk [193].

Now, let us think of the following problem: we have a hypercube as defined in Def. 6.1.3 and we are interested in measuring the time (or, equivalently, the number of steps) an algorithm would take to go from node i to node j, i.e. its *hitting time* (Def. 4.2.3). Since defining the notion of hitting time for a quantum walk is not straightforward, [157] has proposed the following definitions.

Definition 6.1.4.   One-shot hitting time. *A quantum walk U has a (T, p) one-shot $(|\phi_0\rangle, |x\rangle)$ hitting time if the probability to measure state $|x\rangle$ at time T starting in $|\phi\rangle_0$ is larger than p, i.e. $||\langle x|U^T|\phi_0\rangle||^2 \geq p$.*

Definition 6.1.5. $|x\rangle$-stopped walk. *A $|x\rangle$-stopped walk from U starting in state $|\phi_0\rangle$ is the process defined as the iteration of a measurement with the two projectors $\hat{\Pi}_0 = \hat{\Pi}_x = |x\rangle\langle x|$ and $\hat{\Pi}_1 = \hat{I} - \hat{\Pi}_0$. If $\hat{\Pi}_1$ is measured, an application of U follows. If $\hat{\Pi}_0$ is measured the process is stopped.*

Definition 6.1.6.   Concurrent hitting time. *A quantum walk U has a (T, p) concurrent $(|\phi_0\rangle, |x\rangle)$ hitting time if the $|x\rangle$-stopped walk from U and initial state $|\phi_0\rangle$ has a probability $\geq p$ of stopping at a time $t \leq T$.*

In both cases (Defs. 6.1.4 and 6.1.6), it was shown in [157] that the hitting time from one corner to its opposite is polynomial. However, although it was thought that this polynomial hitting time would imply an exponential speedup over corresponding classical algorithms, that is not the case as it is possible to build a polynomial-time classical algorithm to traverse the hypercube from one corner to its opposite [19]. Further studies on hitting times of quantum walks on graphs can be found in [194–196].

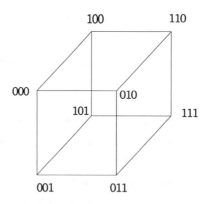

FIGURE 6.1: A three-dimensional hypercube. Nodes are labeled following the formula $d \oplus e_d$ where $d \in \{000, 001, 010, 011, 100, 101, 110, 111\}$ and $e_d \in \{001, 010, 100\}$.

A natural step further along employing discrete quantum walks for solving search problems is to use quantum computation techniques to find items stored in spaces of two or more dimensions. In [197], Benioff proposed the use of Grover's algorithm for searching items in a grid of $\sqrt{N} \times \sqrt{N}$ elements, and showed that a direct application of such algorithm would take $O(N)$ time steps to find one item, i.e. there would be no more quantum speedup. Later on, in [198], Aaronson and Ambainis used Grover's algorithm and multilevel recursion to build algorithms capable of searching in a two-dimensional grid in $O(\sqrt{N}\log^2 N)$ steps and a three-dimensional grid in $O(\sqrt{N})$ steps. Moreover, Childs and Goldstone [199] developed a continuous quantum walk algorithm to solve the search problem in a grid and discovered algorithms that would have an optimal performance of $O(\sqrt{N})$ in grids of five or more dimensions.

Ambainis et al. [200] proposed algorithms based on discrete quantum walks (evolution operators used in [200] are those "perturbed" operators defined in [156]) that would take $O(\sqrt{N}\log N)$ steps to search in a two-dimensional grid and would reach an optimal performance of $O(\sqrt{N})$ for three- and higher dimensional grids. An important contribution of [200] was to show that the performance of search algorithms based on quantum walks is sensitive to the selection of coin operators, i.e. the performance of a search algorithm may be optimal or not depending on the coin operator choice. Finally, Aaronson and Ambainis have shown in [201] how to build algorithms based on discrete quantum walks to search on a two-dimensional grid using a total number of $O(\sqrt{N}\log^{5/2} N)$ steps, and a three-dimensional grid with $O(\sqrt{N})$ number of steps.

A variant of Def. 6.1.2, the **element distinctness problem**, was analyzed in [202].

Definition 6.1.7.   Element distinctness problem *[37]. Given a list of strings over {0, 1} separated by #, determine if all the strings are different.*

A quantum algorithm for solving the element distinctness problem is given in [202]. This algorithm combines the quantum search of spatial regions proposed in [201] with a quantum walk.

The first part of [202] transforms the string list from Def. 6.1.7 into a graph G with marked and non-marked vertices; in this process, [202] uses an oracle (Def. 6.1.1). The second part of the algorithm employs a discrete quantum walk to search graph G. As a result, the algorithm solves the distinctness problem in a total number of $O(N^{2/3})$ steps and $O(N^{\frac{k}{k+1}})$ steps for k identical strings, among N items. Upon the work presented in [202], Magniez et al. proposed in [203] a new quantum algorithm for solving the *triangle problem*, which can be stated as follows.

Definition 6.1.8.   *Let G be a graph. Any complete subgraph of G on three vertices is called a triangle. The triangle problem (in oracle version) can be posed as follows:*

Oracle input: the adjacency matrix f of a graph G on n nodes.
Oracle output: a triangle if there is any, otherwise reject.

Additionally, another quantum algorithm, based on Grover's search quantum algorithm [18], is presented in [203] for solving the same triangle problem.

One more application of [202] has been proposed by Childs and Eisenberg in [204], where it has been proposed to employ the quantum algorithm developed for the distinctness problem (Def. 6.1.7) to solve the L-subset finding (oracle) problem, which can be stated as follows.

Definition 6.1.9. The triangle problem (oracle version).
Oracle input: (1) A black box function $f : D \to R$, where D, R are finite sets and $|D| = n$ is the problem size. (2) Property $P \subset (D \times R)^L$.
Oracle output: Some subset $L = \{x_1, \dots, x_L\} \subset D$ such that $((x_1, f(x_1)), \dots, (x_L, f(x_L)) \in P$, or reject if none exists.

An alternative and refreshing approach to discrete quantum walks is presented by Szegedy [205], where a new definition of a discrete quantum walk in presented via the quantization of a stochastic matrix, as well as an alternative definition of hitting time for discrete quantum walks. Reference [205] begins by defining the search problem as follows.

Definition 6.1.10.   Search problem via stochastic processes. *Given a Markov chain with transition probability matrix $P = (p_{x,y})$ on a discrete state space X, with $|X| = n$, u a given probability distribution on X, and a subset of marked elements $M \subseteq X$, compute an estimate for the number t of iterations required to find an element of M, assuming that the Markov chain is started from a u–distributed element of X.*

Reference [205] continues by defining the following concepts.

Definition 6.1.11. *P_M is the matrix obtained from P by deleting its rows and columns indexed from M.*

Since there is no "natural" (i.e. straightforward) method for quantizing a discrete Markov chain, [205] proposes a quantization method of P which uses bipartite random walks.

Definition 6.1.12. *Let X and Y be two finite sets and $P = (p_{x,y})$ and $Q = (q_{y,x})$ be matrices describing probabilistic maps $X \to Y$ and $Y \to X$, respectively. If we have a single probabilistic function P from X to X, i.e. a Markov chain, in order to create a bipartite walk we can set $q_{y,x} = p_{x,y}$ for every $x, y \in X$ (that is, we set $Q = P$).*

The quantization method for (P, Q) proposed by Szegedy is as follows. We start by creating two operators on the Hilbert space with basis states $|x\rangle$, $|y\rangle$, where $x \in X$ and $y \in Y$. Let us define the states

$$\phi_x = \sum_{y \in Y} \sqrt{p_{x,y}} |x\rangle |y\rangle, \tag{6.4a}$$

$$\psi_y = \sum_{x \in X} \sqrt{q_{y,x}} |x\rangle |y\rangle \tag{6.4b}$$

for every $x \in X$, $y \in Y$. Finally, let us define $A = (\phi_x)$ as the matrix composed of column vectors ϕ_x ($x \in X$), and $B = (\psi_y)$ as the matrix composed of column vectors ψ_y ($y \in Y$). Then, [205] defines the unitary operator W, the quantization of the bipartite walk (P, Q), as

Definition 6.1.13. $W = (2AA^* - I)(2BB^* - I)$.

Szegedy [205] proceeds to build definitions and theorems for new quantum hitting time and upper bounds for finding a marked element as in Def. 6.1.10. A remarkable feature of [205] is a proposal for a new link between classical and quantum walks, namely the development of a quantum walk evolution operator W via a classical stochastic matrix P.

Finally, upon the quantum walk definition given in [205], Magniez et al. [206] proposed a quantum-walk-based algorithm for solving the following problem.

Theorem 1.   *[206] Let $\delta > 0$ be the eigenvalue gap of a reversible, ergodic Markov chain P, and let $\epsilon > 0$ be a lower bound on the probability that an element chosen from the stationary distribution of P is marked whenever M is non-empty. Then, there is a quantum algorithm that high probability determines if M is empty or finds an element of M, with cost of order $S + \frac{1}{\sqrt{\epsilon}}(\frac{1}{\sqrt{\delta}}U + C)$, where S is the computational cost of constructing superposition states, and U, C are costs of constructing unitary transformations as defined in [206, p. 2].*

A summary of quantum search algorithms can be found in [109], and a review of algorithmic applications of quantum walks can be found in [207]. Also, Kendon [111] surveys the algorithmic properties of quantum walks and analyzes several relevant properties of quantum walks like their quantumness and the impact of decoherence in algorithm performance.

Finally, a novel application of discrete quantum walks is shown in [208], where a quantum algorithm for combinatorial optimization problems is proposed. This quantum algorithm combines techniques from discrete quantum walks, quantum phase estimation, and quantum Zeno effect, and can be seen as a quantum counterpart of classical simulated annealing based on Markov chains.

6.1.2 Algorithms Based on Continuous Quantum Walks

The operation and mathematical formulation of discrete quantum walks fit very well into the mindset of a computer scientist, as time evolves in discrete steps (as a typical classical algorithm would) and the model employs walkers and coins, usual elements of stochastic processes when employed in algorithm development. However, the most successful applications of quantum walks are found within the realm of continuous quantum walks, as we shall see in this section. Given the seminal result derived by Strauch [108] about the connection between discrete and continuous quantum walks, we now know that results from continuous quantum walks should be translatable, at least in principle, to discrete quantum walks and vice versa.

Nonetheless, the mathematical structure of continuous quantum walks and the physical meaning of corresponding equations provide an accurate picture of several physical systems upon which we may implement quantum walks and quantum computers. Although most of the physical implementations in this field have been based on the discrete quantum walk model [172, 209–213], the additional stimulus provided by [108] as well as the computational universality of quantum walks [214] and recent connections found between quantum walks and adiabatic quantum computation [105], another model of continuous quantum computation, it is reasonable to expect new implementations based on continuous quantum walks.

Consequently, and taking into account that one of the aims of quantum computing is to harness the physical properties of Nature to compute, it is the opinion of this author that computer scientists and students of computer science should become increasingly knowledgeable about the mathematics of continuous systems and their physical interpretation. Introductory references on this matter can be found in [27, 45] and, for a more advanced treatment of time-dependent quantum mechanics, we refer the reader to [215].

Exponential Algorithmic Speedup by a Quantum Walk

Farhi and Gutmann [182] introduced an algorithm based on a continuous quantum walk, i.e. a quantum walk whose evolution in time is *not* given in discrete steps, but it rather evolves continuously in time according to the Schrödinger equation (Eq. (2.10)).

The proposed algorithm solves the following problem: given a graph G_s consisting of two balanced binary trees of height n with 2^n leaves of the left tree identified with the 2^n leaves of the right tree according to the way shown in Fig. 6.2(a), and with two marked nodes *ENTRANCE* and *EXIT*, find an algorithm to go from *ENTRANCE* to *EXIT*.

It was shown in [182] that it is possible to build a quantum walk that traverses graph G_s from *ENTRANCE* to *EXIT* which is exponentially faster than its corresponding classical random walk [167]. In other words, the *hitting time* of the continuous quantum walk proposed

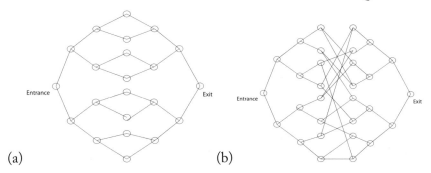

(a)   (b)

FIGURE 6.2: Balanced and unbalanced trees.

in [182] is of polynomial order, while the hitting time of the corresponding classical random walk is of exponential order. However, this advantage does not lead to an exponential speedup due to the fact that it is possible to build a deterministic algorithm that traverses the same graph in polynomial time [19].

Ideas from [182] were taken one step further by Childs et al. [19], where the authors introduced a more general type of graphs G_r to be crossed, proved that those graphs could not be passed across efficiently with any classical algorithm, and delivered an algorithm based on a continuous quantum walk that traverses the graph in polynomial time.

Graphs G_r are built as follows. Begin by constructing two balanced binary trees of height n (i.e. with 2^n leaves), but instead of identifying the leaves, they are connected by a random cycle that alternates between the leaves of the two trees, that is, we choose a leaf on the left at random and connect it to a leaf on the right chosen at random too. Then, we connect the latter to a leaf on the left chosen randomly among the remaining ones. The process is continued, always alternating sides, until every leaf on the left is connected to two leaves on the right, and vice versa. See Fig. 6.2(b) for an example of graphs G_r.

In order to build the quantum walk that will be used to traverse a graph G_r, the authors of [19] defined a Hamiltonian $\hat{H}$ based on the graph adjacency matrix A (Def. 4.2.1). $\hat{H}$ has matrix elements given by

$$\langle a|\hat{H}|a\rangle = \begin{cases} \gamma, & a \neq a', aa' \in G_r \\ 0, & \text{otherwise.} \end{cases} \tag{6.5}$$

In the continuous quantum walk algorithm proposed in [19], the authors used an oracle to learn about the structure of the graph G_r, i.e. information about the Hamiltonian given by Eq. (6.5) is extracted using an oracle. By doing so, it is proved in [19] that it is possible to construct a continuous quantum walk that would traverse any graph G_r in polynomial time. An improved lower bound for any classical algorithm traversing G_r has been proposed in [216], but the performance difference between quantum and classical algorithms in [19] remains exponential.

Let us present a final reflexion with respect to algorithms purely based on quantum walks, before proceeding to talk about other applications of continuous quantum walks. As rightly argued by Ritcher [217], the quantum algorithms reviewed in this chapter are instances of an abstract search problem: given a state space which can be translated into a graph structure, find a marked state (or set of states) by performing a quantum walk on the graph. With this abstraction in mind as well as with the purpose of combining the power of quantum walks with classical sampling algorithms, Ritcher [217] has proposed a method for almost-uniform sampling based on repeated measurements of a continuous quantum walk.

Simulation of Quantum Systems

One of the main goals of quantum computation is the simulation of quantum systems, i.e. the realization of programmable quantum systems whose physical properties allow us to model the behavior of other quantum systems [6, 9, 21].

A novel use of continuous quantum walks for simulation of quantum processes has been presented by Mohseni et al. [218]. In this contribution, the authors have developed a theoretical framework for studying quantum interference effects in energy transfer phenomena, with the purpose of modeling photosynthetic processes. The main contribution of [218] is to analyze the action of the environment in the coherent dynamics of quantum systems related to photosynthesis. The framework developed in [218] includes a generalization of a non-unitary continuous quantum walk in a directed graph (as opposed to a previous definition of a unitary continuous quantum walk on undirected graphs [19]).

6.2   UNIVERSALITY OF QUANTUM WALKS

Universality is a desirable property of a model of computation because it shows that such a model is capable of simulating any other model of computation. Basically, models of computation that are labeled as universal are capable of solving the same problems (although it could happen in different time regimes).

The history of quantum computing includes the recollection of significant efforts to prove the universality of several models of quantum computers, i.e. that any algorithm that can be computed by a general-purpose quantum computer [66] can also be executed by quantum gates [3, 4], as well as computers based on the quantum adiabatic theorem [219–221], for example.

It has recently been proposed that quantum walks are universal models of quantum computation [105, 214]. This result, together with the computational equivalence proofs of several other models of quantum computations, provide a rich "toolbox" for computer scientists interested in quantum computation, for they will be free to choose from several models of quantum computation those that particularly suit their academic background and interests. However, this freedom must be balanced with a profound knowledge about discrete and continuous models of quantum computation, due to the viability of experimental realizations of quantum computers.

CHAPTER 7

Conclusions

In this lecture we have focused on an emerging field of quantum computation: quantum walks. In order to provide a solid background and to situate this field in an appropriate context, we have produced several introductory chapters covering introductions to the fields of quantum mechanics, the theory of computation, and classical discrete random walks.

In our chapter on quantum mechanics we have discussed the postulates of quantum mechanics used to study quantum walks and, more generally, the basic concepts of quantum computation. This chapter is meant to be used by computer scientists and other practitioners from scientific and engineering areas interested in a concise presentation of those concepts of quantum mechanics needed to understand our fields.

As for the theory of computation, we have reviewed the roots of two of the main contributions of Alan Turing to the science of computation: the Church–Turing thesis and Turing machines, together with those elements of the theory of complexity needed to measure the performance of algorithms. These measures are used to quantify the performance of both classical and quantum algorithms. We have also studied the deterministic and nondeterministic models of computation. Finally, we have given a short introduction to the ideas that have enriched the dialogue between physics and computation and provided a formal definition of a Quantum Turing Machine. We have worked on this chapter having in mind not only computer scientists interested in reviewing fundamental elements of computer science, but also physicists looking for a succinct source of information about those basic concepts of the theory of computation needed to start their journey in the study of quantum walks and quantum computation.

In our chapter on classical random walks we have reviewed the main concepts and theorems used in the application of classical random walks in algorithm development. We have pointed out the fact that if we are to compare the properties of classical and quantum walks on infinite and countable spaces, we need to propose new methods for quantifying performance measures of classical random walks, such as mixing time. Again, we have written this chapter having in mind practitioners of several fields, interested in having a concise source on classical random walks relevant to the study of quantum walks and their algorithmic applications. This is particularly useful because most books on Markov chains are *not* focused on those elements used in algorithmic development. Finally, we provide a concise introduction to classical continuous random walks.

In our chapter on quantum walks we present a comprehensive review of the state of the art in discrete and continuous quantum walks. In addition to a careful analysis of quantum walks on a line with Hadamard and arbitrary coin operators, we provide a detailed study of the advantages of the Hadamard quantum walk on the line over its classical counterparts.

We then proceed to review several concepts and theorems on discrete quantum walks on Cayley graphs. We have shown in this chapter that there are several measures (not necessarily equivalent to each other) used to quantify the performance of quantum walks on graphs. This suggests that there is a clear need to produce better performance definitions in order to gain a deeper understanding of the nature of quantum walks on graphs (not only on Cayley graphs but also any other kind of graph that may be useful in algorithm development).

The second part of our chapter on quantum walk starts by addressing connections between classical and quantum walks. We then provide a concise presentation of continuous quantum walks, and present some questions about the quantumness of quantum walks. We finish this chapter by focusing on the randomness of quantum walks, the mathematical transformations that connect discrete and continuous quantum walks, and the relevance (or essential need) of using coins in discrete quantum walks.

The penultimate chapter of this lecture focuses on algorithmic applications of discrete and continuous quantum walk. We present several algorithms based on discrete and continuous quantum walks which have significant advantages over their classical counterparts. In particular, we review an algorithm based on a continuous quantum walk that provides an exponential speedup with respect to its classical counterparts. Then, we present a proposal for the simulation of photosynthetic processes using a generalization of a continuous quantum walk, and finish this chapter with a brief review of the computational universality of quantum walks.

It is common wisdom among historians that the division of history into periods of time is quite an arbitrary procedure [222, 223]. Thus, I would like to state that the division I propose in the following lines is based on my perception of the increasing impact of quantum walks in the scientific community at large.

The first part of the history of quantum walks is composed of all the seminal works developed by a small core of physicists, mathematicians, and computer scientists, who devoted themselves to provide the foundations of our discipline, namely to depict the quantum-mechanical systems whose behavior could be modeled as a walk in a lattice. This era includes the early works of Feynman [104], Aharonov [106], Meyer [112], Nayak and Vishwanath [113], as well as the generalizations of quantum walks on graphs and the very first quantum algorithms based on quantum walks presented in Chapters 5 and 6, respectively.

In the second and current part of the history of quantum walks, research workers filled the gaps and answered some of the fundamental questions produced during the early years of our discipline. I would say that this second era starts with the very encouraging algorithmic results published in [19], followed by the connection between discrete and quantum walks delivered in [108], the proposals about computational universality of quantum walks [105, 214], and, finally, the employment of quantum walks in the simulation of quantum phenomena [218].

A brief study of the past is a good vehicle for designing our future. In my opinion, the future of quantum walks and quantum computing as a whole is very much linked to finding problems of both scientific merit and practical relevance for which quantum algorithms provide better solutions than their classical counterparts, as well as to the successful employment of quantum walks in the simulation of complex physical systems. The mathematical, physical, and computational properties of quantum walks make this field an appealing subject for research workers in order to employ it as a computational tool and/or a physical testbed for scientific discovery. I particularly think of using quantum walks in problems coming from molecular biology and medicine, due to the scientific challenges and the enormous social interest found in those fields.

However, a disadvantage of quantum walks is the somewhat long time that must be invested in the corresponding learning curve. In order to use quantum walks for both algorithm development and quantum simulation, computer scientists must include an additional "toolbox" in their curricula, namely the fundamental properties of reversible gates, the basic mathematical and physical ideas behind quantum mechanics, and the limitations imposed by physical implementations. However, it is indeed the opinion of this author that the educational investment required to learn the field of quantum walks will prove to be fruitful, as the computer science community will enable itself to harness several properties of Nature for computational purposes.

References

[1] J. Brown, *The Quest for the Quantum Computer*. New York: Touchstone, 2001.

[2] J. Volpi, *In Search of Klingsor*. London: Fourth Estate Ltd, 2004.

[3] M. A. Nielsen and I. L. Chuang, *Quantum Computation and Quantum Information*. Cambridge: Cambridge University Press, 2000.

[4] A. Y. Kitaev, A. H. Shen, and M. N. Vyhalyi, *Classical and Quantum Computation*. (Graduate Studies in Mathematics 47), Providence, RI: American Mathematical Society, 1999.

[5] R. P. Feynman, "Simulating physics with computers," *Int. J. Theor. Phys.*, Vol. 21(6/7), pp. 467–488, 1982. doi:10.1007/BF02650179

[6] I. Kassal, S. P. Jordan, P. J. Love, M. Mohseni, and A. Aspuru-Guzik, "Quantum algorithms for the simulation of chemical dynamics," *arXiv:0801.2986*, 2008.

[7] A. Perdomo, C. Truncik, I. Tubert-Brohman, G. Rose, and A. Aspuru-Guzik, "On the construction of model Hamiltonians for adiabatic quantum computation and its application to finding low energy conformations of lattice protein models," *Phys. Rev. A*, Vol. 78, p. 012320, 2008. doi:10.1103/PhysRevA.78.012320

[8] Advanced Research and Development Activity, "Qist: a quantum information science and technology roadmap," 2004.

[9] ERA-Pilot, "Quantum information processing and communication strategic report version 1.4," 2007.

[10] J. Joo, Y. L. Lim, A. Beige, and P. L. Knight, "Single-qubit rotations in 2d optical lattices with multi-qubit addressing," *Phys. Rev. A*, Vol. 74, p. 042344, 2006. doi:10.1103/PhysRevA.74.042344

[11] B. P. Lanyon, T. J. Weinhold, N. K. Langford, M. Barbieri, D. F. V. James, A. Gilchrist, and A. G. White, "Experimental demonstration of a compiled version of Shor's algorithm with quantum entanglement," *Phys. Rev. Lett.*, Vol. 99, p. 250505, 2007. doi:10.1103/PhysRevLett.99.250505

[12] R. Prevedel, P. Walther, F. Tiefenbacher, P. Böhi, R. Kaltenbaek, T. Jennewein, and A. Zeilinger, "High-speed linear optics quantum computing using active feed-forward," *Nature*, Vol. 445, pp. 65–69, 2007. doi:10.1038/nature05346

[13] M. P. A. Branderhorst, P. Londero, P. Wasylczyk, C. Brif, R. L. Kosut, H. Rabitz, and I. A. Walmsley, "Coherent control of decoherence," *Science*, Vol. 320(5876), pp. 638–643, 2008. doi:10.1126/science.1154576

[14] D. Porras and J. I. Cirac, "Quantum manipulation of trapped ions in two dimensional Coulomb crystals," *Phys. Rev. Lett.*, Vol. 96, p. 250501, 2006. doi:10.1103/PhysRevLett.96.250501

[15] J. Benhelm, G. Kirchmair, C. F. Roos, and R. Blatt, "Towards fault-tolerant quantum computing with trapped ions," *Nat. Phys.*, Vol. 4, p. 463, 2008. doi:10.1038/nphys961

[16] D. Deutsch and R. Josza, "Rapid solutions of problems by quantum computation," *Proc. R. Soc. Lond.* A, Vol. 439, pp. 553–558, 1992.

[17] P. W. Shor, "Polynomial-time algorithms for prime factorization and discrete algorithms on a quantum computer," *SIAM J. Comput.*, Vol. 26(5), pp. 1484–1509, 1997. doi:10.1137/S0097539795293172

[18] L. K. Grover, "A fast quantum mechanical algorithm for database search," in *Proc. 28th Annu. ACM Symp. on the Theory of Computing*, 1996, pp. 212–219.

[19] A. M. Childs, R. Cleve, E. Deotto, E. Farhi, S. Gutmann, and D. Spielman, "Exponential algorithmic speedup by quantum walk," in *Proc. 35th ACM Symp. on the Theory of Computation (STOC'03)*, 2003, pp. 59–68. doi:10.1145/780542.780552

[20] D. Horn and A. Gottlieb, "Algorithm for data clustering in pattern recognition problems based on quantum mechanics," *Phys. Rev. Lett.*, Vol. 88, p. 18702, 2002. doi:10.1103/PhysRevLett.88.018702

[21] A. Aspuru-Guzik, A. D. Dutoi, P. J. Love, and M. Head-Gordon, "Simulated quantum computation of molecular energies," *Science*, Vol. 309(5741), pp. 1704–1707, 2005. doi:10.1126/science.1113479

[22] D. Bouwmeester, A. Ekert, and A. Zeilinger, Eds., *The Physics of Quantum Information*. Berlin: Springer, 2001.

[23] R. Motwani and P. Raghavan, *Randomized Algorithms*. Cambridge: Cambridge University Press, 1995.

[24] U. Schöning, "A probabilistic algorithm for k-sat and constraint satisfaction problems," in *Proc. 40th Annu. Symp. on Foundations of Computer Science (FOCS), IEEE*, 1999, pp. 410–414.

[25] W. H. Press, S. A. Teukolsky, W. T. Vetterling, and B. P. Flannery, *Numerical Recipes in C*. Cambridge: Cambridge University Press, 2002.

[26] R. P. Feynman, *Feynman Lectures on Computation*. Baltimore, MD: Penguin Books, 1999.

[27] R. P. Feynman, R. B. Leighton, and M. Sands, *The Feynman Lectures on Physics*, Vol. III. Reading, MA: Addison-Wesley, 1965.

[28] J. Gruska, *Quantum Computing*. New York: McGraw-Hill, 1999.

[29] S. Imre and F. Balázs, *Quantum Computing and Communications: An Engineering Approach*. New York: Wiley and Sons, Ltd, 2005.

[30] V. Vedral, *Introduction to Quantum Information Science*. Oxford: Oxford University Press, 2006.

[31] S. Loepp and W. K. Wootters, *Protecting Information: From Classical Error Correction to Quantum Cryptography*. Cambridge: Cambridge University Press, 2006.

[32] N. D. Mermin, "From cbits to qbits: teaching computer scientists quantum mechanics," *Am. J. Phys.*, Vol. 71, pp. 23–30, 2003. doi:10.1119/1.1522741

[33] E. Rieffel and W. Polak, "An introduction to quantum computing for non-physicists," *ACM Comput. Surv.*, Vol. 32(3), pp. 300–335, 2000. doi:10.1145/367701.367709

[34] B. J. Copeland, *The Essential Turing*. Oxford: Oxford University Press, 2004.

[35] C. H. Papadimitriou, *Computational Complexity*. Reading, MA: Addison-Wesley, 1995.

[36] C. H. Papadimitriou and H. Lewis, *Elements of the Theory of Computation*. Englewood Cliffs, NJ: Prentice-Hall, 1982.

[37] M. Sipser, *Introduction to the Theory of Computation*. Boston, MA: PWS Publishing Co., 2005.

[38] R. Coleman, *Stochastic Processes*. London: George Allen & Unwin, Ltd, 1974.

[39] C. M. Grinstread and J. L. Snell, *Introduction to Probability*. Providence, RI: American Mathematical Society, 1997.

[40] W. Woess, *Random Walks on Infinite Graphs and Groups*. Cambridge Tracts in Mathematics, Vol. 138, Cambridge: Cambridge University Press, 2000.

[41] L. Lovász, "Random walks on graphs: a survey," in *Combinatorics, Paul Erdös is Eighty*, Vol. 2, D. Miklós, V. T. Sós, and T. Szönyi, Eds. Budapest: János Bolyai Mathematical Society, 1996, pp. 353–398.

[42] L. Lovász and P. Winkler, "Mixing times," in *Microsurveys in Discrete Probability*. D. Aldous and J. Propp, Eds., DIMACS Series in Discrete Math. and Theor. Comput. Sci., Providence, RI: American Mathematical Society, 1998, pp. 85–133.

[43] H. Rantanen, "Analyzing the random-walk algorithm for SAT," Master's thesis, Helsinki University of Technology, 2004.

[44] J. Kempe, "Quantum random walks—an introductory overview," *Contemp. Phys.*, Vol. 44(4), pp. 307–327, 2003. doi:10.1080/00107151031000110776

[45] C. Cohen-Tannoudji, B. Diu, and F. Laloe, *Quantum Mechanics*, Vols. 1 and 2. New York: Wiley-Interscience, 1977.

[46] A. Einstein, *Ideas and Opinions*. New York: Wing Books, 1954.

[47] W. Heisenberg, *Physics and Philosophy*. Baltimore, MD: Penguin Group, 1962.

[48] D. Preston, *Before the Fall-out: From Marie Curie to Hiroshima*. New York: Doubleday, 2005.

[49] J. A. Wheeler and W. H. Zurek, Eds., *Quantum Theory and Measurement*. Princeton, NJ: Princeton University Press, 1983.

[50] E. Galvão, "Foundations of quantum theory and quantum information applications," D.Phil. thesis, Centre for Quantum Computation, University of Oxford, 2002.

[51] J. S. Bell, *Speakable and Unspeakable in Quantum Mechanics*. Cambridge: Cambridge University Press, 1987.

[52] P. A. M. Dirac, *The Principles of Quantum Mechanics*. Oxford: Oxford University Press, 1930.

[53] T. Apostol, *Calculus*, Vol. 2. New York: Wiley and Sons, Ltd, 1967.

[54] J. I. Cirac, "Entanglement and distillability," *Notes of the International Summer School 2002*. Instituto Superior Tecnico, Lisbon, Portugal, 2002.

[55] W. K. Wootters and W. H. Zurek, "A single quantum cannot be cloned," *Nature*, Vol. 299, pp. 802–803, 1982. doi:10.1038/299802a0

[56] D. Dieks, "Communication by epr devices," *Phys. Lett.* A, Vol. 92(6), pp. 271–272, 1982. doi:10.1016/0375-9601(82)90084-6

[57] N. J. Cerf and J. Fiurasek, "Optical quantum cloning: a review," *Prog. Opt.*, Vol. 49, p. 455 (Ed. E. Wolf), Elsevier, 2006.

[58] C. H. Bennett, H. J. Bernstein, S. Popescu, and B. Schumacher, "Concentrating partial entanglement by local operations," *Phys. Rev.* A, Vol. 53, pp. 2046–2052, 1996. doi:10.1103/PhysRevA.53.2046

[59] A. Einstein, B. Podolsky, and N. Rosen, "Can quantum mechanical description of physical reality be considered complete?" *Phys. Rev.*, Vol. 47, pp. 777–780, 1935. doi:10.1103/PhysRev.47.777

[60] M. Seevinck and G. Svetlichny, "Bell-type inequalities for partial separability in n-particle systems and quantum mechanical violations," *Phys. Rev. Lett.*, Vol. 89, p. 060401, 2002. doi:10.1103/PhysRevLett.89.060401

[61] D. Collins, N. Gisin, S. Popescu, D. Roberts, and V. Scarani, "Bell-type inequalities to detect true n-body nonseparability," *Phys. Rev. Lett.*, Vol. 88, p. 170405, 2002. doi:10.1103/PhysRevLett.88.170405

[62] M. R. Garey and D. S. Johnson, *Computers and Intractability: A Guide to the Theory of NP-Completeness*. New York: W. H. Freeman and Co., 1979.

[63] R. Zach, "Hilbert's program," In Edward N. Zalta, Ed. *The Stanford Encyclopedia of Philosophy*. Fall 2003.

[64] J. J. Gray, *The Hilbert Challenge*. Oxford: Oxford University Press, 2000.

[65] A. M. Turing, "On computable numbers, with an application to the entscheidung problem," *Proc. Lond. Math. Soc.*, Vol. 42, pp. 230–265, 1936–1937. doi:10.1112/plms/s2-42.1.230

[66] D. Deutsch, "Quantum theory, the Church–Turing principle and the universal quantum computer," *Proc. R. Soc. Lond. Ser. A: Math. Phys. Sci.*, Vol. 400(1818), pp. 97–117, 1985.

[67] S. A. Cook, "The complexity of theorem-proving procedures," in *Proc. 3rd Annu. ACM Symp. on the Theory of Computing*, 1971, pp. 151–158.

[68] I. Gent and T. Walsh, "The search for satisfaction," *Internal Report, Department of Computer Science*, University of Strathclyde, 1999.

[69] P. Bjesse, T. Leonard, and A. Mokkedem, "Finding bugs in an alpha microprocessor using satisfiability solvers," in *Proc. 13th Int. Conf. on Computer Aided Verification*, 2001.

[70] M. N. Velev and R. E. Bryant, "Effective use of boolean satisfiability procedures in the formal verification of superscalar and vliw microprocessors," in *Proc. 38th Design Automation Conf. (DAC '01)*, 2001, pp. 226–231. doi:10.1145/378239.378469

[71] H. Kautz and B. Selman, "The state of sat," *Preliminary Version in Proc. CP-2003. Discrete Appl. Math.* Vol. 155(12), pp. 1514–1524, 2007.

[72] S. A. Cook, "An overview of computational complexity," *Turing Award Lecture*, Association for Computing Machinery, 1983.

[73] L. Fortnow and S. Homer, "A short history of computational complexity," in *The History of Mathematical Logic*. D. van Dalen, J. Dawson, and A. Kanamori, Eds. Amsterdam: North-Holland, 2003.

[74] S. Mertens, "Computational complexity for physicists," *Comput. Sci. Eng. IEEE*, May–June 2002, pp. 31–47.

[75] C. H. Papadimitriou, "On selecting a satisfying truth assignment," in *Proc. 32nd IEEE Symp. on the Foundations of Computer Science*, 1991, pp. 163–169.

[76] J. von Neumann, *Fourth University of Illinois Lecture (Theory of Self-Reproducing Automata)*. Champaign, IL: University of Illinois Press, 1966.

[77] R. Landauer, "Irreversibility and heat generation in the computing process," *IBM J. Res. Dev.*, Vol. 3, p. 183–191, 1961.

[78] C. H. Bennett, "Logical reversibility of computation," *IBM J. Res. Dev.*, Vol. 17, pp. 525–532, 1973.

[79] E. Fredkin and T. Toffoli, "Conservative logic," *Int. J. Theor. Phys.*, Vol. 21, pp. 219–253, 1982. doi:10.1007/BF01857727

[80] Y. Lecerf, "Logique mathématique: machines de Turing réversibles," *C. R. Séances Acad. Sci.*, Vol. 257, pp. 2597–2600, 1963.

[81] P. A. Benioff, "The computer as a physical system: a microscopic quantum mechanical Hamiltonian model of computers as represented by Turing machines," *J. Stat. Phys.*, Vol. 22(5), p. 563, 1980. doi:10.1007/BF01011339

[82] P. A. Benioff, "Quantum mechanical Hamiltonian models of discrete processes that erase their own histories: application to Turing machines," *Int. J. Theor. Phys.*, Vol. 21, pp. 177–201, 1982. doi:10.1007/BF01857725

[83] P. A. Benioff, "Quantum mechanical Hamiltonian models of Turing machines," *J. Stat. Phys.*, Vol. 3(29), pp. 515–546, 1982. doi:10.1007/BF01342185

[84] P. A. Benioff, "Quantum mechanical models of Turing machines that dissipate no energy," *Phys. Rev. Lett.*, Vol. 48, pp. 1581–1585, 1982. doi:10.1103/PhysRevLett.48.1581

[85] A. C. Yao, "Quantum circuit complexity," in *Proc. 34th IEEE Symp. on Foundations of Computer Science*, 1993, pp. 352–361.

[86] E. Bernstein and U. Vazirani, "Quantum complexity theory," *SIAM J. Comput.*, Vol. 5(26), pp. 1411–1473, 1997. doi:10.1137/S0097539796300921

[87] C. S. Calude, J. Casti, and M. J. Dineen, Eds., *Unconventional Models of Computation*. Berlin: Springer, 1998.

[88] G. Grimmett and D. Welsh, *Probability: An Introduction*. Oxford: Oxford University Press, 1991.

[89] J. R. Norris, *Markov Chains*. Cambridge: Cambridge University Press, 1999.

[90] K. F. Riley, M. P. Hobson, and S. J. Bence, *Mathematical Methods for Physics and Engineering*. Cambridge: Cambridge University Press, 1998.

[91] S. M. Ross, *A First Course in Probability*. London: Macmillan Publishing Co., 1984.

[92] P. G. Doyle and J. L. Snell, *Random Walks and Electric Networks* (The Carus Mathematical Monographs 28), Mathematical Association of America, 1984.

[93] G. Pólya, "Über eine aufgabe der wahrscheinlichkeitstheorie betreffend die irrfahrt im straßennetz, English translation: On an exercise in probability concerning the random walk in the road network," *Math. Ann.*, Vol. 84, pp. 149–160, 1921. doi:10.1007/BF01458701

[94] F. Spitzer, *Principles of Random Walk*. Berlin: Springer, 2nd edition, 1976.

[95] J. Kempe, "Calcul Quantique—Marches Aléatoires Quantiques et Etude d'Enchevêtrement," Ph.D. thesis, École Nationale Supérieure de Télécommunications, 2001.

[96] A. Ambainis, E. Bach, A. Nayak, A. Vishwanath, and J. Watrous, "One-dimensional quantum walks," in *Proc. 33rd ACM Symp. on the Theory of Computation (STOC'01) ACM*, 2001, pp. 60–69.

[97] P. Tetali, "Design of on-line algorithms using hitting times," *SIAM J. Comput.*, Vol. 28(4), pp. 1232–1246, 1999. doi:10.1137/S0097539798335511

[98] B. Aspvall, M. F. Plass, and R. E. Tarjan, "A linear-time algorithm for testing the truth of certain quantified boolean formulas," *Inform. Process. Lett.*, Vol. 8(3), pp. 121–123, 1979. doi:10.1016/0020-0190(79)90002-4

[99] K. Iwama and S. Tamaki, "Improved upper bounds for 3-sat," Electronic Colloquium on Computational Complexity, Report 53, 2003.

[100] C. H. Yeang and M. Szummer, "Continuous Markov random walks," in *Proc. 18th Conf. of Uncertainty in Artificial Intelligence (UAI), Acapulco, Mexico*, 2003.

[101] S. Godoy and S. Fujita, "A quantum random-walk model for tunneling diffusion in a 1d lattice," *J. Chem. Phys.*, Vol. 97(7), pp. 5148–5154, 1992. doi:10.1063/1.463812

[102] S. P. Gudder, *Quantum Probability*. San Francisco, CA: Academic, 1988.

[103] N. Konno, "Limit theorems and absorption problems for quantum random walks in one dimension," *Quantum Inform. Comput.*, Vol. 2, pp. 578–595, 2002.

[104] R. P. Feynman, "Quantum mechanical computers," *Found. Phys.*, Vol. 16(6), pp. 507–531, 1986. doi:10.1007/BF01886518

[105] B. A. Chase and A. J. Landhahl, "Universal quantum walks and adiabatic algorithms by 1d Hamiltonians," *arXiv:0802.1207*, 2008.

[106] Y. Aharonov, L. Davidovich, and N. Zagury, "Quantum random walks," *Phys. Rev.* A, Vol. 48, pp. 1687–1690, 1993. doi:10.1103/PhysRevA.48.1687

[107] A. Patel, K. S. Raghunathan, and P. Rungta, "Quantum random walks do not need a coin toss," *Phys. Rev.* A, Vol. 71, p. 032347, 2005. doi:10.1103/PhysRevA.71.032347

[108] F. W. Strauch, "Connecting the discrete and continuous-time quantum walks," *Phys. Rev.* A, Vol. 74, p. 030301, 2006. doi:10.1103/PhysRevA.74.030301

[109] A. Ambainis, "Quantum search algorithms," *SIGACT News*, Vol. 35, pp. 22–35, 2004.

[110] O. Maloyer and V. Kendon, "Decoherence vs entanglement in coined quantum walks," *New J. Phys.*, Vol. 9, p. 87, 2007. doi:10.1088/1367-2630/9/4/087

[111] V. Kendon, "A random walk approach to quantum algorithms," *Philos. Trans. R. Soc.* A, Vol. 364(1849), pp. 3407–3422, 2007.

[112] D. A. Meyer, "From quantum cellular automata to quantum lattice gases," *J. Stat. Phys.*, Vol. 85, pp. 551–574, 1996. doi:10.1007/BF02199356

[113] A. Nayak and A. Vishwanath, "Quantum walk on the line," *quant-ph/0010117*.

[114] A. P. Hines and P. C. E. Stamp, "Quantum walks, quantum gates, and quantum computers," *Phys. Rev.* A, Vol. 75, p. 062321, 2007. doi:10.1103/PhysRevA.75.062321

[115] M. Hamada, N. Konno, and E. Segawa, "Relation between coined quantum walks and quantum cellular automata," *RIMS Kokyuroku*, Vol. 1422(20050400), pp. 1–11, 2005.

[116] N. Konno, K. Mistuda, T. Soshi, and H. J. Yoo, "Quantum walks and reversible cellular automata," *Phys. Lett.* A, Vol. 330(6), pp. 408–417, 2004. doi:10.1016/j.physleta.2004.08.025

[117] W. van Dam, "Quantum cellular automata," M.Sc. thesis, University of Nijmegen, The Netherlands, 1996.

[118] I. Fuss, L. White, P. Sherman, and S. Naguleswaran, "An analytic solution for one-dimensional quantum walks," arXiv:0705.0077v1.

[119] E. Feldman and M. Hillery, "Modifying quantum walks: a scattering theory approach," *J. Phys. A: Math. Theor.*, Vol. 40, pp. 11343–11359, 2007. doi:10.1088/1751-8113/40/37/011

[120] J. Košík, "Two models of quantum random walk," *Central Eur. J. Phys.*, Vol. 4, pp. 556–573, 2003.

[121] H. A. Carteret, M. E. H. Ismail, and B. Richmond, "Three routes to the exact asymptotics for the one-dimensional quantum walk," *J. Phys. A: Math. Gen.*, Vol. 36(33), pp. 8775–8795, 2003. doi:10.1088/0305-4470/36/33/305

[122] H. A. Carteret, B. Richmond, and N. M. Temme, "Evanescence in coined quantum walks," *J. Phys. A: Math. Gen*, Vol. 38, pp. 8641–8665, 2005. doi:10.1088/0305-4470/38/40/011

[123] C. Bender and S. Orszag, *Advanced Mathematical Methods for Scientists and Engineers* (International Series in Pure and Applied Mathematics). New York: McGraw-Hill, 1978.

[124] N. Bleistein and R. Handelsman, *Asymptotic Expansions of Integrals*. New York: Holt, Rinehart and Winston, 1975.

[125] V. Kendon and B. Tregenna, "Decoherence in discrete quantum walks," *Selected Lectures from DICE 2002. Lecture Notes in Physics*, Vol. 633, pp. 253–267, 2003.

[126] N. Konno, "Quantum random walks in one dimension," *Quantum Inform. Process.*, Vol. 1(5), pp. 345–354, 2002. doi:10.1023/A:1023413713008

[127] T. A. Brun, H. A. Carteret, and A. Ambainis, "Quantum walks driven by many coins," *Phys. Rev. A*, Vol. 67, p. 052317, 2003. doi:10.1103/PhysRevA.67.052317

[128] N. Konno, "Symmetry of distribution for the one-dimensional Hadamard walk," *Interdisciplinary Inform. Sci.*, Vol. 10, pp. 11–22, 2004.

[129] N. Konno, "A new type of limit theorems for the one-dimensional quantum random walk," *J. Math. Soc. Japan*, Vol. 57, pp. 1179–1195, 2005. doi:10.2969/jmsj/1150287309

[130] N. Konno, "A path integral approach for disordered quantum walks in one dimension," *Fluctuation Noise Lett.*, Vol. 5(4), pp. 529–537, 2005. doi:10.1142/S0219477505002987

[131] M. Katori, S. Fujino, and N. Konno, "Quantum walks and orbital states of a weyl particle," *Phys. Rev. A*, Vol. 72, p. 012316, 2005. doi:10.1103/PhysRevA.72.012316

[132] C. M. Chandrashekar, R. Srikanth, and R. Laflamme, "Optimizing the discrete quantum walk using a su(2) coin," *Phys. Rev. A*, Vol. 77, p. 032326, 2008. doi:10.1103/PhysRevA.77.032326

[133] B. Tregenna, W. Flanagan, R. Maile, and V. Kendon, "Controlling discrete quantum walks: coins and initial states," *New J. Phys.*, Vol. 5, p. 83, 2003. doi:10.1088/1367-2630/5/1/383

[134] N. Inui and N. Konno, "Localization of multi-state quantum walk in one dimension," *Physica A*, Vol. 353, pp. 133–144, 2005. doi:10.1016/j.physa.2004.12.060

[135] N. Konno N. Inui and E. Segawa, "One-dimensional three-state quantum walk," *Phys. Rev. E*, Vol. 72, p. 056112, 2005. doi:10.1103/PhysRevE.72.026113

[136] S. E. Venegas-Andraca, J. L. Ball, K. Burnett, and S. Bose, "Quantum walks with entangled coins," *New J. Phys.*, Vol. 7, p. 221, 2005. doi:10.1088/1367-2630/7/1/221

[137] P. Ribeiro, P. Milman, and R. Mosseri, "Aperiodic quantum random walks," *Phys. Rev. Lett.*, Vol. 93, p. 190503, 2004. doi:10.1103/PhysRevLett.93.190503

[138] D. D'Alessandro, G. Parlangeli, and F. Albertini, "Non-stationary quantum walks on the cycle," *J. Phys. A: Math. Theor.*, Vol. 40, pp. 14447–14455, 2007. doi:10.1088/1751-8113/40/48/010

[139] P. Feinsilver and J. Kocik, "Krawtchouk matrices from classical and quantum walks," *Contemp. Math.*, Vol. 287, p. 83–96, 2002.

[140] M. C. Bañulus, C. Navarrete, A. Pérez, and E. Roldán, "Quantum walk with a time-dependent coin," *Phys. Rev. A*, Vol. 73, p. 062304, 2006. doi:10.1103/PhysRevA.73.062304

[141] L. Ermann, J. P. Paz, and M. Saraceno, "Decoherence induced by a chaotic environment: a quantum walker with a complex coin," *Phys. Rev. A*, Vol. 73, p. 012302, 2006. doi:10.1103/PhysRevA.73.012302

[142] C. M. Chandrashekar, R. Srikanth, and S. Banerjee, "Symmetries and noise in the quantum walk," *Phys. Rev. A*, Vol. 76, p. 022316, 2007. doi:10.1103/PhysRevA.76.022316

[143] V. Kendon and B. Tregenna, "Decoherence can be useful in quantum walks," *Phys. Rev. A*, Vol. 67, p. 042315, 2003. doi:10.1103/PhysRevA.67.042315

[144] V. Kendon and B. Tregenna, "Decoherence in a quantum walk on the line," in *Proc. QCMC*, 2002.

[145] E. Bach, S. Coppersmith, M. Paz Goldshen, R. Joynt, and J. Watrous, "One-dimensional quantum walks with absorbing boundaries," *J. Comput. Syst. Sci.*, Vol. 69(4), pp. 562–592, 2004. doi:10.1016/j.jcss.2004.03.005

[146] T. Yamasaki, H. Kobayashi, and H. Imai, "Analysis of absorbing times of quantum walks," *Phys. Rev. A*, Vol. 68, p. 012302, 2003. doi:10.1103/PhysRevA.68.012302

[147] N. Konno, T. Namiki, T. Soshi, and A. Sudbury, "Absorption problems for quantum walks in one dimension," *J. Phys. A: Math. Gen.*, Vol. 36(1), pp. 241–253, 2003. doi:10.1088/0305-4470/36/1/316

[148] V. Kendon, "Quantum walks on general graphs," *Int. J. Quantum Inform.*, Vol. 4(5), pp. 791–805, 2006. doi:10.1142/S0219749906002195

[149] T. D. MacKay, S. D. Bartlett, L. T. Stephenson, and B. C. Sanders, "Quantum walks in higher dimensions," *J. Phys. A: Math. Gen.*, Vol. 35, pp. 2745–2753, 2002. doi:10.1088/0305-4470/35/12/304

[150] D. Aharonov, A. Ambainis, J. Kempe, and U. Vazirani, "Quantum walks on graphs," in *Proc. 33rd ACM Symp. on the Theory of Computation (STOC'01)*, 2001, pp. 50–59. doi:10.1145/380752.380758

[151]   N. Fjeldsø, J. Midtdal, and F. Ravndal, "Random walks of a quantum particle on a circle," *J. Phys. A: Math. Gen.*, Vol. 21, pp. 1633–1647, 1988. doi:10.1088/0305-4470/21/7/027

[152]   M. Bednarska, A. Grudka, P. Kurzyński, T. Łuczak, and A. Wójcik, "Quantum walks on cycles," *Phys. Lett.* A, Vol. 317, p. 21, 2003. doi:10.1016/j.physleta.2003.08.023

[153]   M. Bednarska, A. Grudka, P. Kurzyński, T. Łuczak, and A. Wójcik, "Examples of nonuniform limiting distributions for the quantum walk on even cycles," *Int. J. Quantum Inform.*, Vol. 2(4), p. 453, 2004. doi:10.1142/S0219749904000444

[154]   N. Inui, Y. Konishi, N. Konno, and T. Soshi, "Fluctuations of quantum random walks on circles," *Int. J. Quantum Inform.*, Vol. 3(3), pp. 535–550, 2005. doi:10.1142/S0219749905001079

[155]   C. Moore and A. Russell, "Quantum walks on the hypercube," in *Proc. 6th Int. Workshop on Randomization and Approximation Techniques in Computer Science (RANDOM'02)*, 2483 of LNCS, 2002, pp. 164–178. doi:10.1007/3-540-45726-7_14

[156]   N. Shenvi, J. Kempe, and R. B. Whaley, "A quantum random walk search algorithm," *Phys. Rev.* A, Vol. 67(5), p. 052307, 2003. doi:10.1103/PhysRevA.67.052307

[157]   J. Kempe, "Discrete quantum walks hit exponentially faster," in *Proc. 7th Int. Workshop on Randomization and Approximation Techniques in Computer Science (RANDOM'03)*, Vol. 2003, pp. 354–369.

[158]   N. Inui, Y. Konishi, and N. Konno, "Localization of two-dimensional quantum walks," *Phys. Rev.* A, Vol. 69, p. 052323, 2004. doi:10.1103/PhysRevA.69.052323

[159]   A. D. Gottlieb, S. Janson, and P. F. Scudo, "Convergence of coined quantum walks in $\mathbb{R}^d$," *Inf. Dim. Anal. Quantum Probab. Rel. Topics*, Vol. 8, p. 129, 2005 (quant-ph/0406072). doi:10.1142/S0219025705001895

[160]   E. Feldman and M. Hillery, "Scattering theory and discrete-time quantum walks," *Phys. Lett.* A, Vol. 324, p. 277, 2004. doi:10.1016/j.physleta.2004.03.005

[161]   O. López-Acevedo and T. Gobron, "Quantum walks on cayley graphs," *J. Phys. A: Math. Gen.*, Vol. 39, pp. 585–599, 2006. doi:10.1088/0305-4470/39/3/011

[162]   A. Montanaro, "Quantum walks on directed graphs," *Quantum Inform. Comput.*, Vol. 7(1), pp. 093–102, 2007.

[163]   H. Krovi and T. A. Brun, "Quantum walks on quotient graphs," *Phys. Rev.* A, Vol. 75, p. 062332, 2007. doi:10.1103/PhysRevA.75.062332

[164]   H. Krovi, "Symmetry in Quantum Walks," Ph.D. thesis, University of Southern California, 2007.

[165]   J. Watrous, "Quantum simulations of classical random walks and undirected graph connectivity," *J. Comput. Syst. Sci.*, Vol. 62(2), pp. 376–391, 2001. doi:10.1006/jcss.2000.1732

[166] T. A. Brun, H. A. Carteret, and A. Ambainis, "Quantum to classical transition for random walks," *Phys. Rev. Lett.*, Vol. 91, p. 130602, 2003. doi:10.1103/PhysRevLett.91.130602

[167] A. Childs, E. Farhi, and S. Gutmann, "An example of the difference between quantum and classical random walks," *Quantum Inform. Process.*, Vol. 1, pp. 35–43, 2002. doi:10.1023/A:1019609420309

[168] G. Alagić and A. Russell, "Decoherence in quantum walks on the hypercube," *Phys. Rev.* A, Vol. 72, p. 062304, 2005. doi:10.1103/PhysRevA.72.062304

[169] J. Košík, V. Bužek, and M. Hillery, "Quantum walks with random phase shifts," *Phys. Rev.* A, Vol. 74, p. 022310, 2006. doi:10.1103/PhysRevA.74.022310

[170] H. Jeong, M. Paternostro, and M. S. Kim, *Phys. Rev.* A, Vol. 69, p. 012310, 2004. doi:10.1103/PhysRevA.69.012310

[171] P. L. Knight, E. Roldán, and J. E. Sipe, "Quantum walk on the line as an interference phenomenon," *Phys. Rev.* A, Vol. 68, p. 020301, 2003. doi:10.1103/PhysRevA.68.020301

[172] P. L. Knight, E. Roldán, and J. E. Sipe, "Optical cavity implementations of the quantum walk," *Opt. Commun.*, Vol. 227, pp. 147–157, 2003. doi:10.1016/j.optcom.2003.09.024

[173] P. L. Knight, E. Roldán, and J. E. Sipe, "Propagating quantum walks: the origin of interference structures," *J. Mod. Opt.*, Vol. 51(12), pp. 1761–1777, 2004.

[174] V. Kendon and B. C. Sanders, "Complementarity and quantum walks," *Phys. Rev.* A, Vol. 71, p. 022307, 2005. doi:10.1103/PhysRevA.71.022307

[175] W. K. Wootters and W. H. Zurek, "Complementarity in the double-slit experiment: quantum nonseparability and a quantitative statement of Bohr's principle," *Phys. Rev.* D, Vol. 19, pp. 473–484, 1979.

[176] A. Romanelli, A. C. Sicardi Schifino, R. Siri, G. Abal, A. Auyuanet, and R. Donangelo, "Quantum random walk on the line as a Markovian process," *Physica* A, Vol. 338(3–4), pp. 395–405, 2004. doi:10.1016/j.physa.2004.02.061

[177] I. Carneiro, M. Loo, X. Xu, M. Girerd, V. Kendon, and P. L. Knight, "Entanglement in coined quantum walks on regular graphs," *New J. Phys.*, Vol. 7, p. 156, 2005. doi:10.1088/1367-2630/7/1/156

[178] G. Abal, R. Siri, A. Romanelli, and R. Donangelo, "Quantum walk on the line: entanglement and non-local initial conditions," *Phys. Rev.* A, Vol. 73, p. 042302, 2006. doi:10.1103/PhysRevA.73.042302

[179] R. Donangelo G. Abal and H. Fort, "Long-time entanglement in the quantum walk," *arXiv:quant-ph/0709.3279*, 2007.

[180] S. E. Venegas-Andraca, "Discrete quantum walks and quantum image processing," D.Phil. thesis, Centre for Quantum Computation, University of Oxford, 2006.

[181] Y. Omar, N. Paunković, L. Sheridan, and S. Bose, "Quantum walk on a line with two entangled particles," *Phys. Rev. A*, Vol. 74, p. 042304, 2006. doi:10.1103/PhysRevA.74.042304

[182] E. Farhi and S. Gutmann, "Quantum computation and decision trees," *Phys. Rev. A*, Vol. 58, pp. 915–928, 1998. doi:10.1103/PhysRevA.58.915

[183] N. Konno, "Continuous-time quantum walks on ultrametric spaces," *Int. J. Quantum Inform.*, Vol. 4(6), pp. 1023–1036, 2006. doi:10.1142/S0219749906002389

[184] N. Konno, "Continuous-time quantum walks on trees in quantum probability theory," *Infinite Dimens. Anal. Quantum Probab. Relat. Top.*, Vol. 9(2), pp. 287–297, 2006. doi:10.1142/S0219025706002354

[185] D. de Falco and D. Tamascelli, "Speed and entropy of an interacting continuous time quantum walk," *J. Phys. A: Math. Gen.*, Vol. 39, pp. 5873–5895, 2006. doi:10.1088/0305-4470/39/20/016

[186] O. Mülken, V. Pernice, and A. Blumen, "Quantum transport on small-world networks: a continuous-time quantum walk approach," *Phys. Rev. E*, Vol. 76, p. 051125, 2007. doi:10.1103/PhysRevE.76.051125

[187] M. A. Jafarizadeh, S. Salimi, and R. Sufiani, "Investigation of continuous-time quantum walk by using krylov subspace-lanczos algorithm," *Eur. Phys. J. B*, 59(2), pp. 199–216, 2007.

[188] S. Abramsky, "A structural apporach to reversible computation," *Theor. Comput. Sci.*, Vol. 347, pp. 441–464, 2005. doi:10.1016/j.tcs.2005.07.002

[189] J. L. Gómez-Muñoz, "Quantum©, a Mathematica© add-on for simulating quantum walks and quantum algorithms in general," Quantum Information Processing Group, Tecnológico de Monterrey Campus Estado de México. http://homepage.cem.itesm.mx/lgomez/quantum/index.htm, 2008.

[190] List of QC Simulators. http://www.quantiki.org/wiki/index.php/doi:wiki/index.php/.

[191] E. Kashefi, A. Kent, V. Vedral, and K. Banaszek, "A comparison of quantum oracles," *Phys. Rev. A*, Vol. 65, p. 050304, 2002. doi:10.1103/PhysRevA.65.050304

[192] A. Gbris, T. Kiss, and I. Jex, "Scattering quantum random-walk search with errors," *Phys. Rev. A*, Vol. 76, p. 062315, 2007. doi:10.1103/PhysRevA.76.062315

[193] M. Hillery, J. Bergou, and E. Feldman, "Quantum walks based on an interferometric analogy," *Phys. Rev. A*, Vol. 68, p. 032314, 2003. doi:10.1103/PhysRevA.68.032314

[194] J. Košík and V. Bužek, "Scattering model for quantum random walks on hypercube," *Phys. Rev. A*, Vol. 71, p. 012306, 2005. doi:10.1103/PhysRevA.71.012306

[195] H. Krovi and T. Brun, "Hitting time for quantum walks on the hypercube," *Phys. Rev. A*, Vol. 73, p. 032341, 2006. doi:10.1103/PhysRevA.73.032341

[196] H. Krovi and T. Brun, "Quantum walks with infinite hitting times," *Phys. Rev.* A, Vol. 74, p. 042334, 2006. doi:10.1103/PhysRevA.74.042334

[197] P. A. Benioff, "Space searches with a quantum robot," in *Quantum Computation and Quantum Information: A Millennium Volume.* S. Lomonaco and H. E. Brandt, Eds., AMS Contemporary Mathematics (305), 2002.

[198] S. Aaronson and A. Ambainis, "Quantum search of spatial regions," in *Proc. 44th Annu. IEEE Symp. on Foundations of Computer Science*, Vol. 2003, pp. 200–209. doi:10.1109/SFCS.2003.1238194

[199] A. M. Childs and J. Goldstone, "Spatial search by quantum walk," *Phys. Rev.* A, Vol. 70, p. 022314, 2004. doi:10.1103/PhysRevA.70.022314

[200] A. Ambainis, J. Kempe, and A. Rivosh, "Coins make quantum walks faster," in *Proc. 16th ACM-SIAM SODA*, Vol. 2005, pp. 1099–1108.

[201] S. Aaronson and A. Ambainis, "Quantum search of spatial regions," *Theor. Comput.*, Vol. 1, pp. 47–79, 2005.

[202] A. Ambainis, "Quantum walk algorithm for element distinctness," in *Proc. 45th Annu. IEEE Symp. on Foundations of Computer Science*, Vol. 2004, pp. 22–31. doi:10.1109/FOCS.2004.54

[203] F. Magniez, M. Santha, and M. Szegedy, "Quantum algorithms for the triangle problem," *SIAM J. Comput.*, Vol. 37(2), pp. 413–424, 2007. doi:10.1137/050643684

[204] A. Childs and J. M. Eisenberg, "Quantum algorithms for subset finding," *Quantum Inform. Comput.*, Vol. 5, p. 593, 2005.

[205] M. Szegedy, "Quantum speed-up of Markov chain algorithms," in *Proc. 45th IEEE Symp. on the Foundations of Computer Science*, Vol. 2004, pp. 32–41. doi:10.1109/FOCS.2004.53

[206] F. Magniez, A. Nayak, J. Roland, and M. Santha, "Search via quantum walk," in *Proc. 39th ACM Symp. on Theory of Computing*, 2007.

[207] A. Ambainis, "Quantum walks and their algorithmic applications," *Int. J. Quantum Inform.*, Vol. 1(4), pp. 507–518, 2003. doi:10.1142/S0219749903000383

[208] R. D. Somma, S. Boixo, and H. Barnum, "Quantum simulated annealing," *arXiv: 0712.1008*, 2007.

[209] C.M. Chandrashekar, "Implementing the one-dimensional quantum (Hadamard) walk using a Bose–Einstein condensate," *Phys. Rev.* A, Vol. 74, p. 032307, 2006. doi:10.1103/PhysRevA.74.032307

[210] C.A. Ryan, M. Laforest, J.C. Boileau, and R. Laflamme, "Experimental implementation of discrete time quantum random walk on an nmr quantum information processor," *Phys. Rev.* A, Vol. 72, p. 062317, 2005. doi:10.1103/PhysRevA.72.062317

[211] E. Roldán and J. C. Soriano, "Optical implementability of the two-dimensional quantum walk," *J. Mod. Opt.*, Vol. 52, pp. 2649–2657, 2005. doi:10.1080/09500340500309873

[212] J. Du, H. Li, X. Xu, M. Shi, J. Wu, X. Zhou, and R. Han, "Experimental implementation of the quantum random-walk algorithm," *Phys. Rev.* A, Vol. 67, p. 042316, 2003. doi:10.1103/PhysRevA.67.042316

[213] B. C. Travaglione and G.J. Milburn, "Implementing the quantum random walk," *Phys. Rev.* A, Vol. 65, p. 032310, 2002. doi:10.1103/PhysRevA.65.032310

[214] A. M. Childs, "Universal computation by quantum walk," *arXiv:0806.1972*, 2008.

[215] D. J. Tannor, *Introduction to Quantum Mechanics: A Time-dependent Perspective.* Mill Valley, CA: University Science Books, 2007.

[216] S. A. Fenner and Y. Zhang, "A note on the classical lower bound for a quantum walk algorithm," *quant-ph/0312230*, 2003.

[217] P. C. Ritcher, "Almost uniform sampling via quantum walks," *New J. Phys.*, Vol. 97 2, 2007.

[218] M. Mohseni, P. Rebentrost, S. Lloyd, and A. Aspuru-Guzik, "Environment-assisted quantum walks in energy transfer of photosynthetic complexes," *arXiv:0805.2741*, 2008.

[219] A. Messiah, *Quantum Mechanics.* New York: Dover, 1999.

[220] E. Farhi, J. Goldstone, S. Gutmann, and M. Sipser, "Quantum computation by adiabatic evolution," *arXiv:quant-ph/0001106*, 2000.

[221] D. Aharonov, W. van Dam, J. Kempe, Z. Landau, S. Lloyd, and O. Regev, "Adiabatic quantum computation is equivalent to standard quantum computation," *SIAM J. Comput.*, Vol. 37(1), pp. 166–194, 2007. doi:10.1137/S0097539705447323

[222] E. H. Carr, *What is History?: The George Macaulay Trevelyan Lectures Delivered in the University of Cambridge.* Baltimore, MD: Penguin Books, 1961.

[223] M. Hughes-Warrington, *Fifty Key Thinkers on History.* London: Routledge, 2000.

Author Biography

Salvador E. Venegas-Andraca is an assistant professor of mathematics and computer science, and head of the Quantum Information Processing group at Tecnologico de Monterrey Campus Estado de Mexico. Dr Venegas-Andraca's research interests include the algorithmic properties of quantum walks, the development of quantum algorithms for solving classical problems and simulating quantum systems, as well as the understanding and application of non-conventional models of computation.

Dr Venegas-Andraca is particularly interested in understanding how to develop quantum and classical strategies for solving problems from the field of molecular biology, particularly protein folding.

Dr Venegas-Andraca holds a BSc in Computer Science and Digital Electronics from Tecnologico de Monterrey, as well as MSc and PhD degrees in Computer Science and Quantum Computation respectively, both degrees from the University of Oxford. He is a true devotee of science, computer technology, history, philosophy and politics. More information on Dr Venegas-Andraca's interests can be found in his web page http://mindsofmexico.org/sva/

Printed in the United States
by Baker & Taylor Publisher Services